INCORPOREAL
IDÉE FIXE

By

Damon Dion Reed

Table of Contents

Chapter 1: Incunabulum

It is silly to think that history means something. If it weren't for the malfeasance of social rejects, we would live much simpler lives without the fear of repeating history. But, we have been enlightened to the past and we are decisively our own people. Therefore, in as much as society has spread out across this globe to avoid the constraints of prior generations, men and women have pushed forward to avoid the constraints of their parents' mental constructs. Maybe progress is genetically engineered into us or maybe it is a relic of the Ego imparted to us via imperfect child rearing. Whatever the case maybe, science was bound to happen. Granted, 'science' didn't yield anything productive for almost two thousand years, but that is beside the point. The point is, we have stepped past the 'ether' believing stage, partaken of the vaccines, and hold our shit together with Slim Jims and Vicodin. And yet, we search endlessly for an ethereal meaning within the minutia of science...and this book is no different.

Over the course of several books, I've postulated a mildly functional unified theory around the concept of Quanta Dynamics, which is based upon the conservation of energy via the synergistic association of continually degrading matter, i.e. energy. But, I would be remise if I didn't warn you about the trillions of unique UNIMAGINED environments that exist outside the pages of any book. Or in simpler terms, the general trends of science are difficult to see when inspecting the minutia of unique elements within unique environments.

The origin of this book is slightly different because instead of flipping through pages and pages of notes, I've taken to lining my walls with ideas. And in as much as it a blatant indicator of my-misdeeds for anyone who dares enter my room, supposedly, it is extremely helpful in identifying patterns. So without further ado, let's look for God's mist about the repetitive minutia of Quanta Dynamics, which has forged our existence.

Chapter 2: Neuproz-glyphics

To begin this arduous journey, let's review some Neuproz notation. The notation of [N] signifies a Neuproz group, which is a proton and a neutron exchanging a negative Neuprotron. The numeral **within** the brackets indicates the type of Neuproz group and a numeral <u>outside</u> the brackets means the number of successive Neuproz groups. For example, please peruse the figure below.

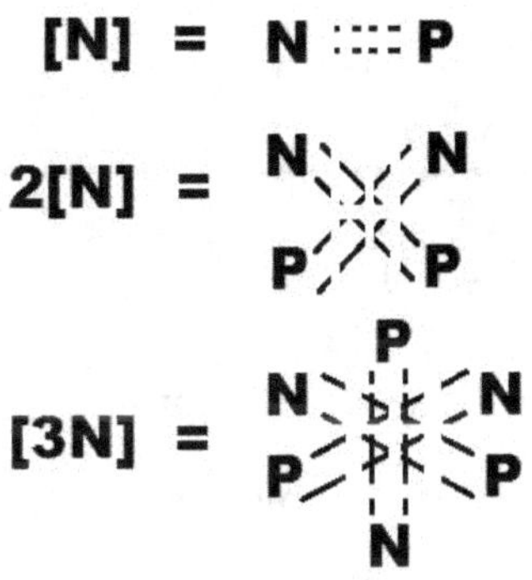

$$[N] \ = \ N \cdots P$$

$$2[N] \ = \ \begin{matrix} N \cdots N \\ \diagdown\diagup \\ \diagup\diagdown \\ P \cdots P \end{matrix}$$

$$[3N] \ = \ \begin{matrix} P \\ N \cdots N \\ P \cdots P \\ N \end{matrix}$$

Figure 1: Neuproz-ology

With this nomenclature in mind, it is important to note that this notation does NOT indicate electronic isolation. It is simply a mathematical tracking system. For example, Nitrogen-14 can be notated as [N]2[3N], but that does NOT mean the [N] grouping is electronically isolated from the other two [3N] groups. Or in scientific terms, the exchange of a negative Neuprotron creates a STRONG nuclear force and the interaction about and between Neuproz groups creates a WEAK columbic nuclear force. As for the distribution of these different Neuproz groupings within elements, it is based upon the 'row' of the periodic table.

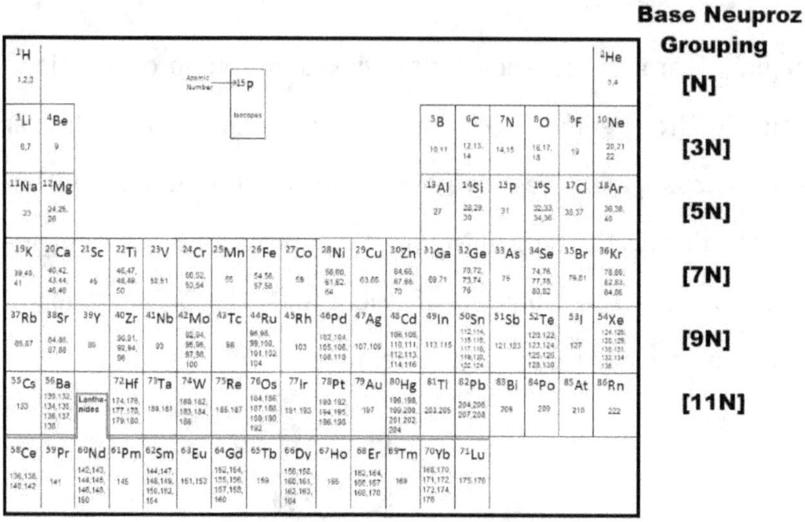

Base Neuproz Grouping

[N]

[3N]

[5N]

[7N]

[9N]

[11N]

Figure 2: Rows are Horizontal[1]

As you can see, each successive row contains a larger Neuproz grouping. But, one unfortunate caveat to this nomenclature is that Noble Gases contain the Neuproz coupling of the previous row, except for Helium...it's too small. (FYI, reference 1 is for the isotopes and NOT for the postulate of Neuproz groups.) For example, Neon-20 exists within the [3N] row and its Neuproz description is as follows: 2[N][3N][5N]. As for the way these Neuproz groupings are spherically orientated, let's look at Fluorine-19's Neuproz Topology created by (N)3[N]2[3N]. (FYI,

6

notice that Fluorine-19 does not have a [5N] Neuproz grouping because it exists in the [3N] row of the periodic table.)

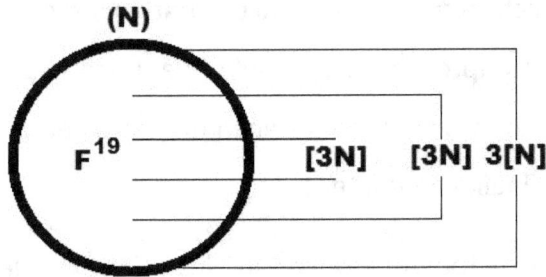

Figure 3: Neuproz Topology

As you can see in this figure, Fluorine-19 has two [3N] groupings about the equator of the atom, three [N] groupings about the poles, and an extra neutron at the top, i.e. (N) notates a neutron NOT a Neuproz grouping. With that review in mind and while we're on the topic of nomenclature, here is some new nomenclature.

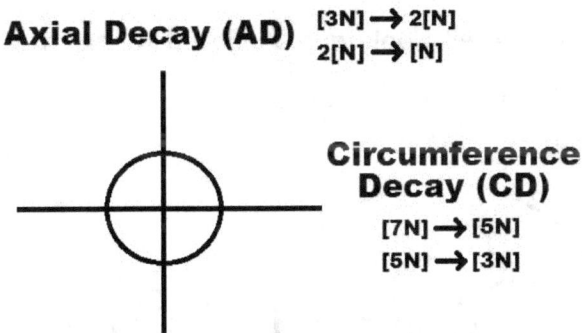

Figure 4: Axial vs. Circumference Decay

As a result of spherically orientated Neuproz groupings, two different types of decays are possible: Axial and Circumference. Generally, Axial Decays release protons, neutrons, and/or negative Neuprotrons. As for

Circumference Decays, they usually release negative Neuprotrons, i.e. the relaxing of larger Neuproz groupings into smaller Neuproz groupings, which results in the incorporation of extra neutrons in between the Neuproz groups. (FYI, the release of neutrons via circumference decays only results when the extra neutrons are stacked oddly, which I'll talk about later.)

While we're on the topic of atomic particles being extruded from atomic nuclei, I need to point out that heavier elements often decay via the release of helium atoms. Unfortunately, since the center of the atomic nucleus is obstructed by the continual sharing of negative Neuprotrons between Neuproz groupings, it is **impossible** that two [N] groups, which are stretched across the circumference of the atom, can axially release a helium atom. Or in simpler terms, the atomic particles of the two [N] groupings **cannot** fly THROUGH the center of the atomic nucleus. All of which, has left me with only one viable postulate to the release of helium atoms from heavier elements.

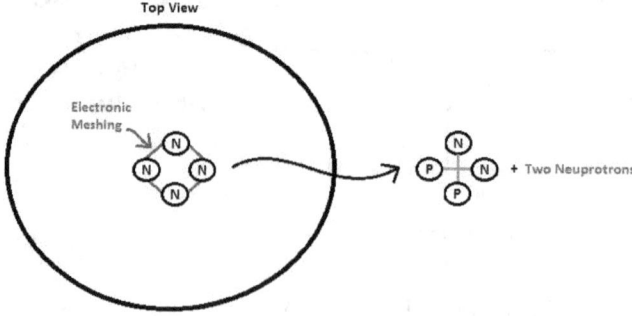

Figure 5: Helium Decay

As you can see in this figure, four included neutrons at the top of the atom forge a new helium atom, which is released axially. As for the reason this is possible? Well, as a result of intense electronic meshing, i.e. weak columbic nuclear forces, the density of atomic particles within heavier elements, and the fact that heavier elements have a ton of extra neutrons, it is possible that four electronically aligned neutrons are simultaneously released such that they forge a helium atom via the loss of two negative Neuprotrons (FYI, more than likely this release is axially, but it is possible that odd neutron stacking can result in a helium circumference decay.)

In conclusion, atomic nuclei are spherically arranged layers of Neuproz groupings with extra neutrons stuck in between the Neuproz groupings. The Neuproz groupings are notated by [brackets], the numeral outside the [brackets] indicates the number of successive Neuproz groupings, and (parenthesizes) notate the extra neutrons, which are held in place by weak columbic nuclear forces. And finally, there is a spectrum of strong nuclear forces {[13N]>[11N]...[3N]>[N]} and weak columbic nuclear forces {[13N]&[11N] > [11N]&[9N] ... [3N]&[N] > [N]&(N)}. (FYI, I will go into more detail about the spectrum of nuclear forces in Chapter 4.)

Chapter 3: Decaying Factors

Traditional elemental decay was postulated based upon the DECTECTION of released energy. But, since we exist in a negative branch of the universe and atoms are surrounded by massive amounts of negative energy, it is surprising we DETECT as much radiation as we do. Or in other terms, radiation DECTECTION is done with extremely negative magnetic environments, which has the propensity to degrade the radiating energy. All of which, is one of the reasons why I postulated that Neuproz pairs are exchanging negative Neuprotrons instead of positive positrons. In any event, let's look at some anomalies caused Quanta Dynamics, i.e. the continual decay of energy.

If you ever get in an argument about the merits of a new scientific field, i.e. Quanta Dynamics...hopefully, here are some things you might want to remember. First, neutrons degrade into protons in about 10 minutes within a laboratory setting[1], which is usually COLDER than liquid hot magma. Next, the periodic table PERIODICALLY has elements with more isotopes. And finally, just do-es it. (FYI, I had to change that for legal reasons, but it worked for Nike.)

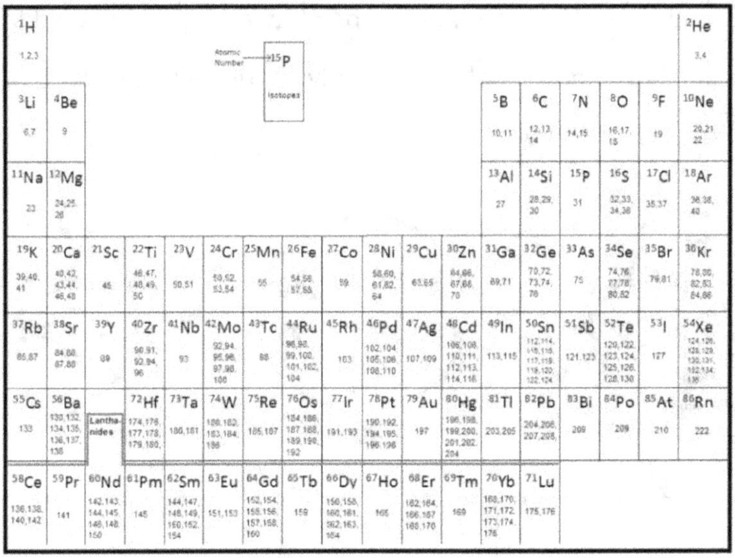

¹H 1,2,3				*Atomic Number* →15p *Isotopes*								⁵B 10,11	⁶C 12,13,14	⁷N 14,15	⁸O 16,17,18	⁹F 19	¹⁰Ne 20,21,22	²He 3,4

Figure 6: The Periodic Table[1]

Is it just me or is it ODD that every other EVEN element has an abundance of isotopes, which are elements that have a variable amount of extra neutrons? On second thought, let me propose a METHOD instead of asking questions.

As a result of intensely negative environments, which enhances the columbic dissonance within the atomic nucleus by enlarging the atomic particles, a proton can degrade into a neutron via the release of a negative Neuprotron.

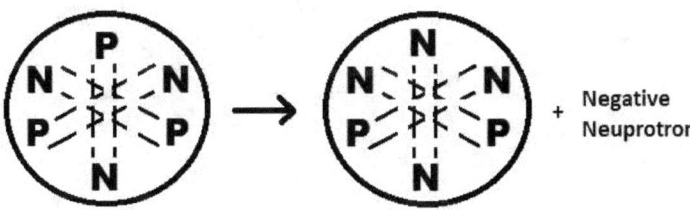

Figure 7: Rate Determining Step

11

Then, sometime later, a number of the neutrons are expunged from the atomic nucleus because they are only being held in place by weak columbic nuclear forces.

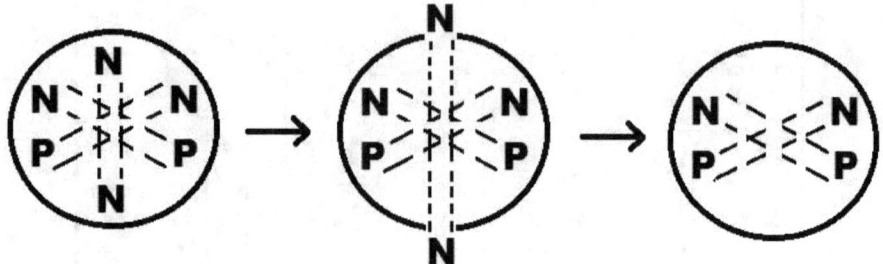

Figure 8: Isotope Conversion

As for the reason why neutron emission is not detected more often? Well, neutrons only have a half-life of about 10 minutes[1] in this negative branch of the universe, i.e. neutrons degrade into protons really-really-REALLY quickly.

Another oddity of the periodic table is that every other isotope is usually more abundant. For example, check out the abundances of these isotopes.

Element	Isotope Weight (Abundance)		
^{19}K	39 (93.2581)	40 (0.0117)	41 (6.7302)
^{12}Mg	24 (78.99)	25 (10.0)	26 (11.01)
^{10}Ne	20 (90.48)	21 (0.27)	22 (9.25)
^{8}O	16 (99.762)	17 (0.038)	18 (0.2)

Table 1: Every Other[1]

As you can see in this figure, ^{18}O, ^{22}Ne, ^{26}Mg, and ^{41}K are ALL more abundant than the <u>preceding</u> isotope. All of which makes me wonder: Is Earth's bimodal magnetosphere stabilizing isotopes that have a **symmetric** distribution of excess neutrons, i.e. every other isotope? (FYI, the answer is more than likely yes.)

In conclusion, if you can't detect it, then it NEVER happened. (It's science's version of the Las Vegas's moto.) On second thought, it's tough to go looking for something you never imagined would be there. All of which, is the reason why theoretical science exits. In any event, I postulate the environment plays an important part in determining the rate & TYPE of element decay. For example, symmetric isotopes are usually more abundant, which suggests Earth's bimodal magnetosphere influences the rate & TYPE of elemental decay. As for the reason why the periodic table PERIODICALLY has more isotopes, you'll just have to keep reading to find out.

Chapter 4: Variable Cohesiveness

I feel silly saying this, but I fear it is NOT completely obvious: Neutral things are relatively negative in a negative branch of the universe. Unfortunately, that cognition brings into question the concept of 'neutrality', which is a distance based concept, i.e. 'your thumb appears to be bigger than a distant skyscraper' concept. Or in simpler terms, a neutron might have more positive charges than a proton, but it appears to be neutral because of the location of its negative charges. All of which, relates to variable cohesiveness as it pertains to weak columbic nuclear forces. Or in other words, charges that are displayed by atomic particles determines the weak columbic nuclear forces between atomic particles.

First and foremost, based upon my postulate that Neuproz pairs are protons exchanging a negative Neuprotron and that mass is a function of positivity within this negative branch of the universe, neutrons have a slightly smaller volume than protons.

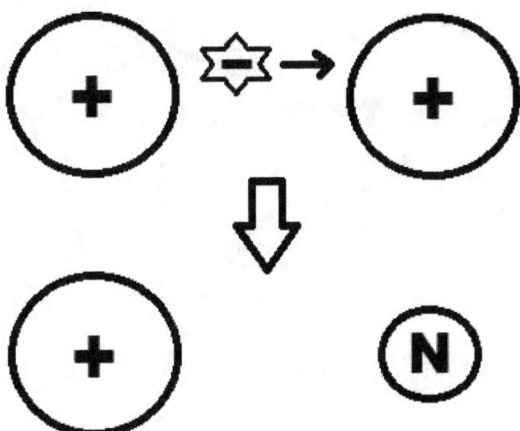

Figure 9: Neuproz Pair

As you can see, the negative Neuprotron star results in the conversion of a proton into a neutron, which is exaggerated to seem smaller in this figure. All of which, leads to the following question: How does a negative Neuprotron innervate with a pair of complex protons such that a rhythmic exchange can be ascertained? Well, the short answer is: A negative Neuprotron star GRINDS around the **columbic grooves** of the spherical proton until the negative Neuprotron can be exchanged, i.e. flung back towards the other proton in the Neuproz pair. And the long answer is, of course, much longer and extremely complicated...please bear with me.

Unfortunately, since the word 'atomic nucleus' is already used to notate the collection of atomic particles within elements, I will use the word 'CORE' to describe the internal electronics within atomic particles. Therefore, let's look at the FUNDAMETNAL basis of the CORE in atomic particles.

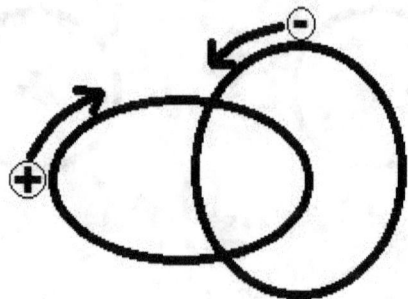

Figure 10: Diffusion Resistant Energy

As you can see in this figure, the interlocked trajectory of opposite charges will prevent diffusion. But, since atomic particles are really small and contain a plethora of moving energetic particles, the CORE is a little bit more complicated.

As any good physicist knows, circles are efficient because they don't have drastic changes in trajectory. And as any good mathematician knows, the infinity symbol is circle with ONE supercoil. And finally, as any biologist knows, supercoiling allows for the storage of 'condensed' DNA within biological atomic nuclei. All of which is displayed below, respectively.

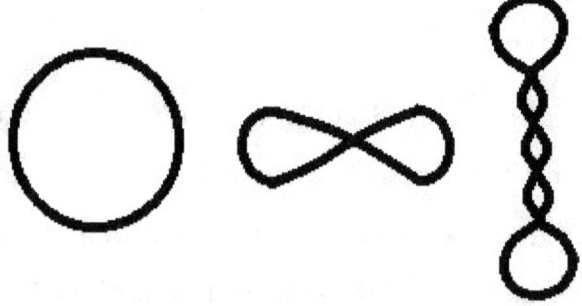

Figure 11: Supercoiling Energy

With all this in mind, I postulate that the CORE of atomic particles have MULTIPLE supercoiled INTERLOCKED energetic quanta, which exists as a 'grooved' sphere such that negative Neuprotron stars can grind around it.

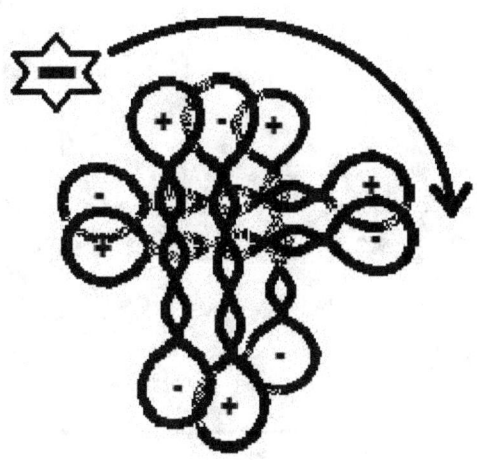

Figure 12: Atomic Particle's CORE

Hopefully, the supercoiled interlocked columbic energy of the CORE APPEARS to be **unsymmetrical** in this figure. (I'll talk about the reason later...hopefully.) What I need to talk about right now is the SPECTRUM of harmonics associated with the exchange of negative Neuprotrons as it relates to a SPECTRUM of strong nuclear forces within different Neuproz groupings. All of which, is related to a spectrum of weak columbic nuclear forces as a function of the atomic particles' external electronics. Quite simply, there are different LAYERS/grooves of negative Neuprotrons grinding about the CORE of atomic particles, which creates a SPECTRUM of strong nuclear forces, external electronics, and weak columbic nuclear forces. Or in specific terms, the linear spectrum of strong nuclear forces is as follows: [11N]>[9N]>[7N]>[5N]>[3N]>[N].

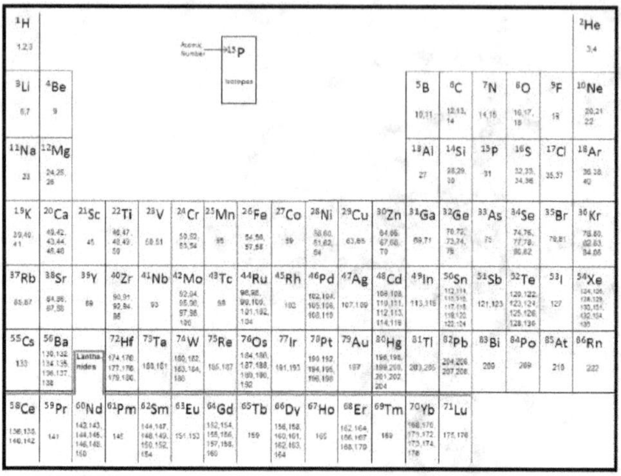

Base Neuproz Grouping	Layers
[N]	1
[3N]	2
[5N]	3
[7N]	4
[9N]	5
[11N]	6

Figure 13: Negative Neuprotron Layers

As you can see in this figure, the [11N] grouping is associated with Layer 6, which is the second to deepest and strongest negative Neuprotron groove. (FYI, I didn't include the last layer of the periodic table, i.e. [13N] = Layer 7.) Or in simpler terms, I'm postulating that the depth/layer of the negative Neuprotron groove relates to the **velocity** of the negative Neuprotron grinding around the CORE. All of which, determines the exchange RATE of the negative Neuprotron between the Neuproz pair, i.e. the strength of the strong nuclear force about Neuproz groupings.

As for the reason why I'm postulating the existence of 7 layers? Well, it is because I like burritos and NOT because there are seven rows to the periodic table. Also, beta decays can convert protons into neutrons and neutrons into protons, which means there are multiple negative Neuprotron layers/grooves. In any event, this groove/layer postulate provides a viable method as to why negative Neuprotrons don't collide about the center of atomic nuclei. Unfortunately, the concept of different negative Neuprotron layers/grooves, i.e. strong nuclear forces, complicates atomic cohesiveness, i.e. weak columbic nuclear forces.

As a result of Layer 3 being a groove that is <u>closer</u> to the CORE of the atomic particle in comparison to the groove of Layer 1, the movement of the negative Neuprotron within Layer 3 creates less variation in the **external** electronic shell of the atomic particle. Or in simpler terms, the movement of a negative Neuprotron in Layer 1 creates a **highly** variable **external** electronic shell to the atomic particle because Layer 1 is the most external groove/layer. Conversely, the movement of a negative Neuprotron in Layer 3 creates LESS variation in the external electronic shell of the atomic particle because Layer 3 is a deeper groove. All of which, allows for more stable columbic interaction, i.e. more cohesion, between atomic particles that are sharing negative Neuprotrons via deeper grooves/layers. (BTW, there is a spectrum of cohesiveness about each layer/groove as it relates to density, time, and temperature...but I won't get into that right now.)

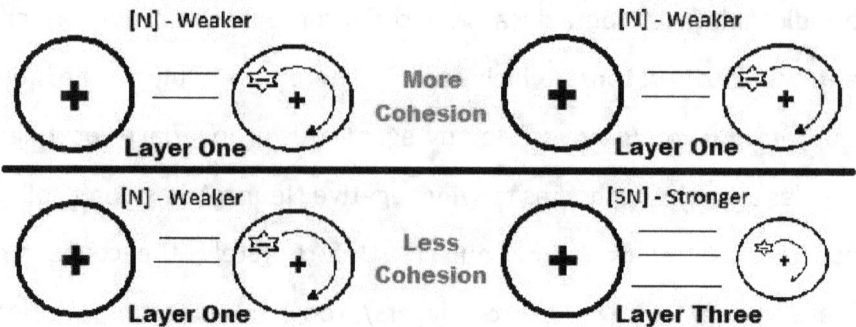

Figure 14: Spectrum of Nuclear Forces

As you can see in the top half of this figure, there is **more cohesion** between Neuproz groupings that are exchanging a negative Neuprotron via Layer 1, which is the result of similar electronics being aligned between the external electronic shells of these atomic particles. But, as you can see in the bottom half of this figure, there is **less cohesion** between the [N] and [5N] grouping because of their dissimilar external electronic shells. Or in simpler terms, the strong nuclear forces within [13N] grouping is greater than in [11N] grouping...and so on and so forth: {[13N]>[11N]...[3N]>[N]}. Also, weak columbic nuclear forces between [13N] and [11N] are stronger than the weak nuclear forces between [11N] and [9N]...and so on and so forth: {[13N]&[11N] > [11N]&[9N]...[3N]&[N] > [N]&(N)}. (FYI, the weak columbic nuclear forces between extra neutrons and Neuproz groupings is dependent which Neuproz grouping the extra neutron originated from.) All of which, is dependent on **environment**.

With all that relative electronic integration based upon negative Neuprotron groove/layer in mind, hopefully it should be apparent why smaller Neuproz groupings degrade first. If not, then let me try and explain. First and foremost, smaller Neuproz groupings exchange negative Neuprotrons via grooves that are more exposed to the environment, i.e. the grooves are further away from the CORE of the atomic particle. Therefore, the intense presence of negative thermal energetic quanta are more likely to disrupt outer grooves, i.e. decrease the electronic cohesion between the negative Neuprotron and the CORE of the atomic particle. Also, the presence of negative thermal energetic quanta causes the expansion of CORE and the 'constriction' of external grooves, which decreases the columbic attraction between the CORE and the negative Neuprotron. In any event, the extreme expansion of axial Neuproz groupings results in the eventual release of a negative Neuprotron star.

Figure 15: Elemental Decay

As you can see in this figure, the expansion of the axial atomic particles causes a proton to release a negative Neuprotron to form a neutron. Then, after the decay of the weak columbic nuclear forces between the neutron and adjacent atomic particles, which is probably facilitated by environment, the neutron is expelled via an Acne Event, i.e. the slow step of elemental decay.

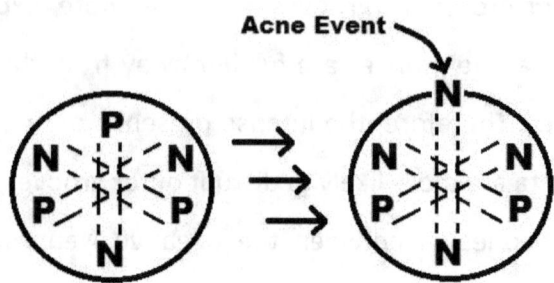

Figure 16: Acne Event

As you can see in this figure, the neutron explodes from the atomic nuclei like a ripe puss laden zit. And even though Acne Events generally occur about the axis of atomic nuclei, the decay of protons into neutrons is quite common about the circumference, which I'll talk about in the next chapter.

In conclusion, I've postulated that larger Neuproz groupings exchange negative Neuprotrons at different rates based upon the layer/groove of negative Neuprotron, i.e. the negative Neuprotrons within [13N] grouping exists about Layer 6 and are exchanged faster than negative Neuprotrons in Layer 5, i.e. [11N] grouping. And as a result of larger Neuproz groupings exchanging negative Neuprotrons via deeper layers, they have more stable external electronic shells, which increases electronic cohesion between the atomic particles in larger Neuproz groups. Next, I postulated that negative thermal energetic quanta result in the expansion/constriction of atomic particles in smaller

22

Neuproz groups, which results in decay via negative Neuprotron emission, i.e. the rate determining step. All of which, results in the **preferential** degradation of smaller Neuproz groups about the axis of the atomic nuclei and lays the framework for the isotopic degradation pathway.

Chapter 5: Isotopic Degradation Pathway

With the concept of 7 different grooves about the CORE of atomic particles as it relates to the rate of negative Neuprotron movement between different Neuproz groups in mind, as well as the expulsion of neutrons via Acne Events, it is time to gaze at a large portion of the periodic table and identify any factors that might lead to trends and subsequently postulates...Damn it, did I just give away my secret to success?

Over the course of the next two pages, I've been able to piece together what I call the Isotopic Degradation Pathway. But, before you take a gander at it, here are some things you might want to know. In the first column is the abbreviation for the element with the atomic number. For those of you who don't know this, the atomic number simple represents the number of protons within the atomic nucleus. In the cells to the right of elemental abbreviation are the weights of all the isotopes as well as their abundance in parenthesis. And finally, the **bolded isotopes (abundance)** decay to produce the <u>underlined</u> isotopes two rows down, which is via the release of two negative Neuprotrons, i.e. two protons decaying into two neutrons. Now in terms of Quanta Dynamics, if you multiply the atomic number by two and subtract that from the weight of the isotope, then you'll know how many extra neutrons there are about the element's Neuproz groupings.

Element	Decay: **Bold** to <u>Underline</u> <Weight (Abundance)>								
82Pb	**204** (1.4)	206 (24.1)	207 (22.1)	208 (52.4)					
81Tl	203 (29.524)	205 (70.476)							
80Hg	196 (0.15)	**198** (9.97)	199 (16.87)	200 (23.10)	201 (13.18)	202 (29.86)	<u>204</u> (6.87)		
79Au	197 (100.0)								
78Pt	190 (0.01)	**192** (0.79)	194 (32.9)	195 (33.8)	196 (25.3)	<u>198</u> (7.2)			
77Ir	191 (37.3)	193 (62.7)							
76Os	184 (0.02)	**186** (1.58)	187 (1.6)	188 (13.3)	189 (16.1)	190 (26.4)	<u>192</u> (41.0)		
75Re	185 (37.40)	187 (62.60)							
74W	**180** (0.13)	182 (26.3)	183 (14.3)	184 (30.67)	<u>186</u> (28.6)				
73Ta	180 (0.012)	181 (99.988)							
72Hf	174 (0.162)	**176** (5.206)	177 (18.606)	178 (27.297)	179 (13.629)	<u>180</u> (35.1)			
71Lu	175 (97.41)	176 (2.59)							
70Yb	168 (0.13)	**170** (3.05)	171 (14.3)	172 (21.9)	173 (16.12)	174 (31.8)	<u>176</u> (12.7)		
69Tm	169 (100.0)								
68Er	162 (0.14)	**164** (1.61)	166 (33.6)	167 (22.95)	168 (26.8)	<u>170</u> (14.9)			
67Ho	165 (100.0)								
66Dy	156 (0.06)	158 (0.10)	**160** (2.34)	161 (18.9)	162 (25.5)	163 (24.9)	<u>164</u> (28.2)		
65Tb	159 (100.0)								
64Gd	**152** (0.20)	154 (2.18)	155 (14.80)	156 (20.47)	157 (15.65)	158 (24.84)	<u>160</u> (21.86)		
63Eu	151 (47.8)	153 (52.2)							
62Sm	**144** (3.1)	147 (15.0)	148 (11.3)	149 (13.8)	150 (7.4)	<u>152</u> (26.7)	154 (22.7)		
61Pm	145 (0.0)								
60Nd	**142** (27.13)	143 (12.18)	<u>144</u> (23.8)	145 (8.3)	146 (17.19)	148 (5.76)	150 (5.64)		
59Pr	141 (100.0)								
58Ce	136 (0.19)	**138** (0.25)	140 (88.48)	<u>142</u> (11.08)					
57La	138 (0.0902)	139 (99.9098)							
56Ba	130 (0.106)	132 (0.101)	**134** (2.417)	135 (6.592)	136 (7.854)	137 (11.23)	<u>138</u> (71.7)		
55Cs	133 (100.0)								
54Xe	124 (0.1)	126 (0.09)	**128** (1.91)	129 (26.4)	130 (4.1)	131 (21.2)	132 (26.9)	<u>134</u> (10.4)	136 (8.9)
53I	127 (100.0)								
52Te	120 (0.096)	**122** (2.603)	123 (0.908)	124 (4.816)	125 (7.139)	126 (18.95)	<u>128</u> (31.69)	130 (33.8)	
51Sb	121 (57.36)	123 (42.64)							
50Sn	112 (0.97)	**114** (0.65)	115 (0.34)	116 (14.53)	117 (7.68)	118 (24.23)	119 (8.59)	120 (32.59)	<u>122</u> (4.63)	124 (5.79)
49In	113 (4.3)	115 (95.7)							
48Cd	106 (1.25)	**108** (0.89)	110 (12.49)	111 (12.8)	112 (24.13)	113 (12.22)	<u>114</u> (28.73)	116 (7.49)	
47Ag	107 (51.839)	109 (48.161)							
46Pd	102 (1.02)	**104** (11.14)	105 (22.33)	106 (27.33)	<u>108</u> (26.46)	110 (11.72)			
45Rh	103 (100.0)								
44Ru	96 (5.52)	**98** (1.88)	99 (12.7)	100 (12.6)	101 (17.0)	102 (31.6)	<u>104</u> (18.7)		
43Tc	98 (0.0)								
42Mo	92 (14.84)	**94** (9.25)	95 (15.92)	96 (16.68)	97 (9.55)	<u>98</u> (24.13)	100 (9.63)		
41Nb	93 (100.0)								
40Zr	**90** (51.45)	91 (11.22)	92 (17.15)	<u>94</u> (17.38)	96 (2.80)				
39Y	89 (100.0)								
38Sr	84 (0.56)	**86** (9.86)	87 (7.0)	<u>88</u> (82.58)					
37Rb	85 (72.165)	87 (27.835)							

Table 2: Isotopic Degradation Pathway Part 1[1]

25

Element	Decay: **Bold** to <u>Underline</u> <Weight (Abundance)>					
³⁶Kr	78 (0.35)	**80** (2.25)	82 (11.6)	83 (11.5)	84 (57.0)	<u>86</u> (17.3)
³⁵Br		79 (50.69)	81 (49.31)			
³⁴Se	74 (0.89)	**76** (9.36)	77 (7.63)	78 (23.78)	<u>80</u> (49.61)	82 (8.73)
³³As		75 (100.0)				
³²Ge	**70** (21.23)	72 (27.66)	73 (7.73)	74 (35.94)	<u>76</u> (7.44)	
³¹Ga	69 (60.108)	71 (39.892)				
³⁰Zn	**64** (48.6)	66 (27.9)	67 (4.1)	68 (18.8)	<u>70</u> (0.6)	
²⁹Cu	63 (69.17)	65 (30.83)				
²⁸Ni	**58** (68.077)	60 (26.223)	61 (1.14)	62 (3.634)	<u>64</u> (0.926)	
²⁷Co		59 (100.0)				
²⁶Fe	**54** (5.8)	56 (91.72)	57 (2.1)	<u>58</u> (0.28)		
²⁵Mn		55 (100.0)				
²⁴Cr	**50** (4.345)	52 (83.789)	53 (9.501)	<u>54</u> (2.365)		
²³V	50 (0.250)	51 (99.750)				
²²Ti	**46** (8.0)	47 (7.3)	48 (73.8)	49 (5.5)	<u>50</u> (5.4)	
²¹Sc	45 (100.0)					
²⁰Ca	**40** (96.941)	42 (0.647)	43 (0.135)	44 (2.086)	<u>46</u> (0.004)	48 (0.187)
¹⁹K	39 (93.2581)	40 (0.0117)	<u>41</u> (6.7302)			
¹⁸Ar	**36** (0.337)	38 (0.063)	<u>40</u> (99.6)			
¹⁷Cl	35 (75.77)	37 (24.23)				
¹⁶S	**32** (95.02)	33 (0.75)	34 (4.21)	<u>36</u> (0.02)		
¹⁵P	31 (100.0)					
¹⁴Si	**28** (92.23)	29 (4.67)	<u>30</u> (3.10)			
¹³Al	27 (100.0)					
¹²Mg	**24** (78.99)	25 (10.0)	<u>26</u> (11.01)			
¹¹Na	23 (100.0)					
¹⁰Ne	**20** (90.48)	21 (0.27)	<u>22</u> (9.25)			
⁹F	19 (100.0)					
⁸O	**16** (99.762)	17 (0.038)	<u>18</u> (0.2)			
⁷N	14 (99.634)	15 (0.366)				
⁶C	**12** (98.90)	13 (1.10)	<u>14</u> (0.0)			
⁵B	10 (19.2)	11 (80.2)				
⁴Be	**<u>9</u>** (100.0)					
³Li	6 (7.5)	<u>**7**</u> (92.5)				

Table 3: Isotopic Degradation Pathway Part 2[1]

For your information, I did not include each element's Neuproz description because it's really annoying to figure out. (BTW, reference 1 refers to the isotopes (abundances) and NOT the postulate of an Isotopic Degradation Pathway.) Also, the placement of the extra neutrons is a BIG deal, which I'll talk about later. For now, let's look at Xenon-128, which has the following Neuproz description: 20(N)[N]2[5N]2[7N]2[9N][11N]. As you can see, hopefully, Xenon-128 has 20 extra neutrons dispersed amongst the array of Neuproz groupings. Now, Xenon-128 degrades into Tellurium-128 via the release of two negative Neuprotrons and by adjusting the Neuproz groupings to: 24(N)[N]2[5N]2[7N]3[9N]. All of which, brings us to the REASON why EVEN elements have more isotopes, i.e. each successive Neuproz downgrade results in a difference of FOUR atomic particles. For example, the [11N] grouping has 11 protons and 11 neutrons, i.e. 22 atomic particles, and the [9N] grouping has 9 protons and 9 neutrons, i.e. 18 atomic particles. Therefore, when [11N] degrades into [9N], it excludes 2 neutrons and 2 protons, which degrade to form two neutrons via the release of two negative Neuprotrons. Or in simpler terms, there are MORE isotopes for EVEN elements because of the difference of atomic particles between successive Neuproz groupings. Unfortunately, things get really complicated when you ask the following question: Where do all the extra neutrons go? The short answer is: Stereo Isotopes. The long answer is much, much longer.

The easiest way to understand Stereo Isotopes is to imagine that Neuproz groupings are the cinderblocks of atomic nuclei and the extra neutrons are the mortar between the cinderblocks, which means Stereo Isotopes occur when the neutron-mortar is about different Neuproz grouping cinderblocks. In any event, please look at the next figure...pretty please?

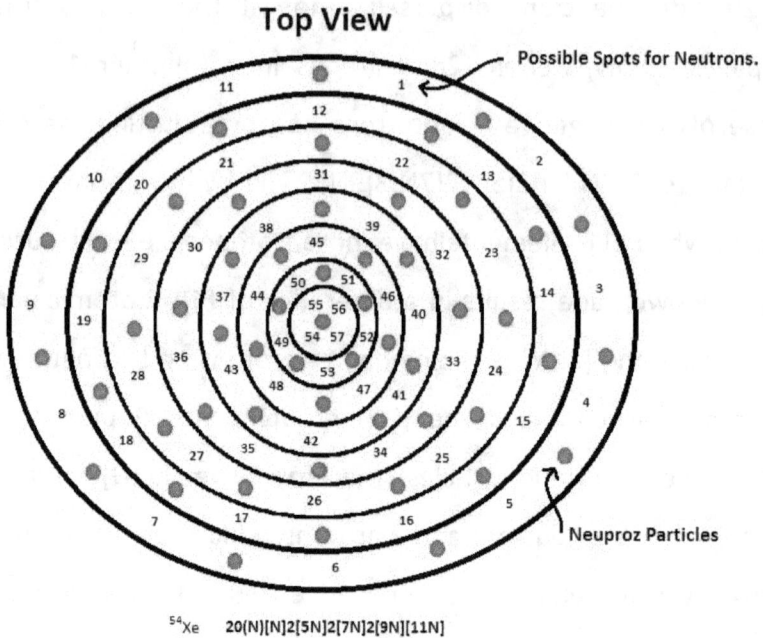

Possible Spots for Neutrons.

Neuproz Particles

^{54}Xe   20(N)[N]2[5N]2[7N]2[9N][11N]

Figure 17: Half the Possibilities

Hopefully you can see that this figure ONLY represents the top half of Xenon-128. Also, I hope you can see that there are **about** 57 POSSIBLE spots for Xenon's 10 extra neutrons between the Neuproz Particles, i.e. the protons & neutrons of Neuproz groupings. (I'll talk about the bolded '**about**' later.) Therefore, as a result of the VARIABLE arrangement of Xenon-128's 20 neutrons about 114 possible spots, there are a multitude of Stereo Isotopes. (FYI, Stereo Isomers is an organic chemistry term used to describe molecules that have the same empirical formula but have a different molecular arrangements.) Or in simpler terms, Stereo Isotopes result when the EXTRA neutrons are placed DIFFERENTLY about the Neuproz groupings. (As for the number of possibilities? I've made it completely clear to all who have read my books that I am NOT a friend to Math. Therefore, I know there are a number of possibilities, but I'm not quite sure as to how many...Damn you statistics!) Another reason I'm apt NOT to employ statistics is because there is bit of variance as to how many neutrons can fit around the Neuproz groups that exist at the circumference and axis of atomic nuclei.

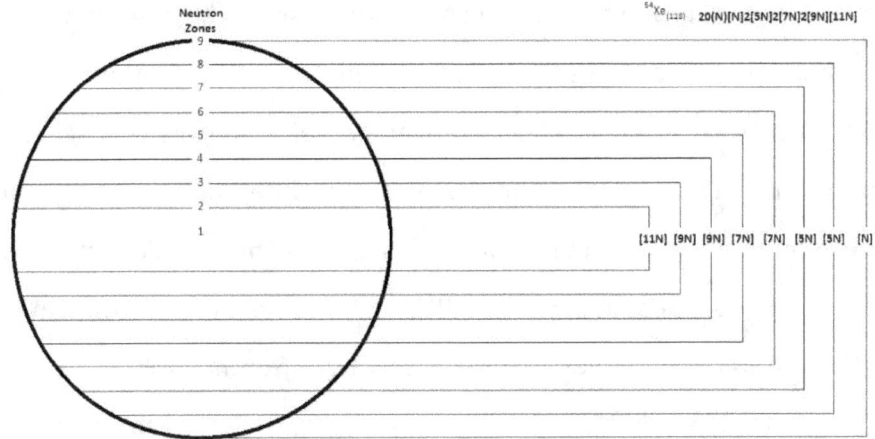

Figure 18: Variance

As you can see in this figure, Xenon-128 has 9 Neutron Zones for EXTRA neutrons. But, the number of neutrons that can go in zone 1, the circumference, varies per the Neuproz grouping about the circumference. Also, the number of neutrons that can go in zone 9, varies per the Neuproz grouping about the axis. Unfortunately, to truly understand the complexity associated with the placement of the extra neutrons, I need to explain the nuance of atomic orbital placement as it relates to isotopic decay as well as the function of different negative Neuprotron layers as it relates to weak columbic nuclear forces and isotopic relaxing.

A little over nine years ago, I had the crazy idea that electrons recharge via traversing the atomic nucleus in search of the positive charge therein. Well, as a result of neutrons being placed in between the Neuproz groupings via the degradation of larger Neuproz groupings into small Neuproz groupings, which are slowly pushing the extra neutrons towards the axis of the atomic nucleus, the UNIQUE placement of the extra neutrons can distort the movement of electrons through the atomic nucleus. And since the movement of different electrons about the atomic nucleus is the POWER that forges the atomic orbitals, it shouldn't be surprising that the VARIABLE placement of neutrons within different Stereo Isotopes results in slightly different atomic orbitals. And just to wrap all this up in a nice package with a cute little bow on top, variable atomic orbital placement about Stereo Isotopes results in DIFFERENT amount plasma about the circumference and/or axis of an atomic nuclei. Or in simpler terms, the unique placement of the extra neutrons about the Neuproz groupings creates unique entrance and/or exit pathways for recharging electrons. All of which, results in DIFFERENT atomic orbitals that allow different amounts of plasma to innervate different regions of the atomic nucleus such that different atomic particles engorge and degrade, i.e. different Stereo Isotopic degradation.

Now, I realize that 'different Stereo Isotopic degradation' sounds as if it goes against the idea of an Isotopic Degradation Pathway, which contained **bolded** isotopes decaying into underlined isotopes two rows down, but the universe is complex. Or in simpler terms, the Isotopic Degradation Pathway represents the MAJOR degradation pathway as it relates to these factors: Abundances, the placement of ODD elements, and the corresponding isotopes two rows down. Or in the simplest terms, the figure you are about to view displays the minutia of feasible degradations based upon the VARIABLE neutron placement in Stereo Isotopes.

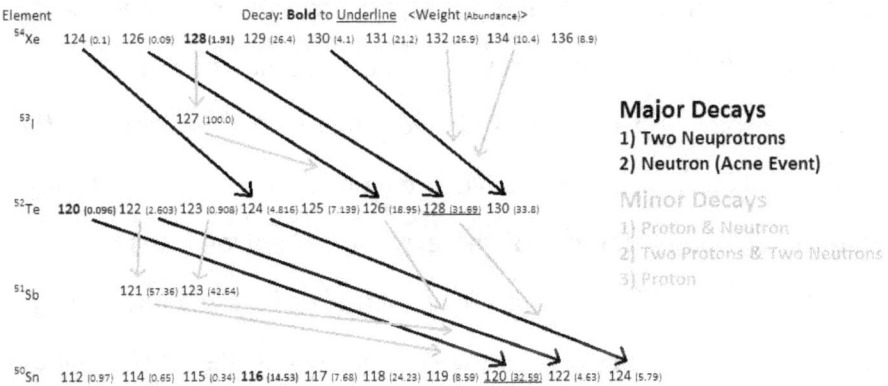

Figure 19: Possibilities

As you can see in this figure, Xenon isotopes 124, 126, 128, and 130 can undergo circumference decays to yield the Tellurium isotopes 124, 126, 128, and 130. Unfortunately, each isotope contains multiple Stereo Isotopes, which can decay differently based upon the placement of the extra neutrons. For example, let's say Xenon 128 has five possible Stereo Isotopes: 128a, 128b, 128c, 128d, and 128e.

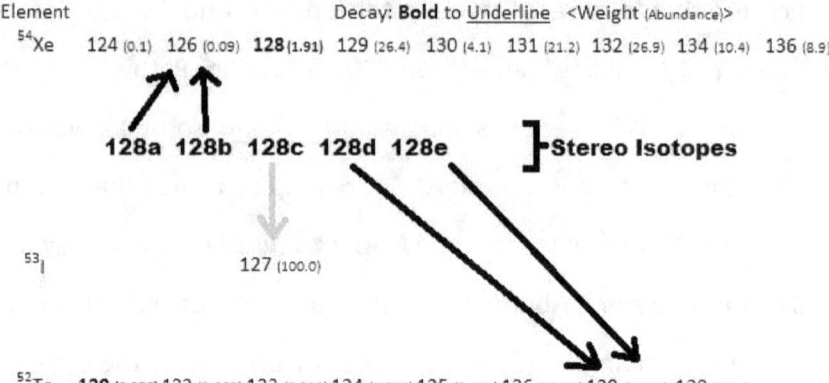

Element   Decay: **Bold** to <u>Underline</u> <Weight (Abundance)>

^{54}Xe   124 (0.1)   126 (0.09)   **128** (1.91)   129 (26.4)   130 (4.1)   131 (21.2)   132 (26.9)   134 (10.4)   136 (8.9)

128a 128b 128c 128d 128e   ]**Stereo Isotopes**

^{53}I   127 (100.0)

^{52}Te   **120** (0.096) 122 (2.603) 123 (0.908) 124 (4.816) 125 (7.139) 126 (18.95) <u>128 (31.69)</u> 130 (33.8)

Figure 20: Stereo Isotope Degradation Possibilities

As you can surmise, the atomic orbitals of Xenon 128d & 128e result in an increase in negative plasma about the circumference of the atomic nuclei, which causes the decay a larger Neuproz group into a small Neuproz group by the release of two negative Neuprotrons to form Tellurium 128, which could have a couple different Stereo Isotopes based upon the placement of the extra four neutrons. As for Xenon 128c, the atomic orbitals results in an increase of plasma about the axis such that it decays via the release of a proton equivalent to form Iodine 127, which could have one or two different Stereo Isotopes. And finally, Xenon 128a & 128b have atomic orbital arrangements that results in plasma about both poles of the atomic nucleus, which results in the expulsion of two neutrons to form Xenon 126, which could result in a couple different Stereo Isotopes.

In the event that your mind is whirling from the spectrum of possibilities of neutron placement as it relates different Stereo Isotopes that have different atomic orbitals based upon different entrances/exists to the atomic nucleus, I might as well hit you with one more complex idea: Weak columbic nuclear forces can cause neutrons to adjust their negative Neuprotrons.

In the Variable Cohesiveness chapter, I postulated that there are 7 different negative Neuprotron Layers/grooves and that Layer 7 is the closest groove to the CORE of the atomic particle. All of which means, DEEPER grooves allow for negative Neuprotrons to grind faster around the CORE without external interference, exchange between Neuproz particles faster, and create a stronger bonds between the Neuproz particles.

With all that in mind, let's imagine that [11N] grouping relaxes into a [9N] grouping via the release of two negative Neuprotrons and the intercalation of four neutrons, which WERE exchanging negative Neuprotrons via Layer 6. All of which means, the intercalated four neutrons each have an empty groove about Layer 6, but they are now surrounded by a [9N] grouping that is exchanging their negative Neuprotrons via Layer/groove 5. As a result of this, do the weak columbic nuclear forces about the [9N] grouping cause the intercalated neutrons to relax, i.e. move a negative Neuprotron from Layer/groove 5 into Layer/groove 6 such that they mimic the electronics of the [9N] grouping? Also, since different Stereo Isotopes have different arrangements of neutrons, does the placement of the extra neutrons determine if it assimilates to the [9N] grouping or a different Neuproz grouping? And finally, are neutrons surrounded ONLY by other neutrons decay faster than neutrons that are assimilated with adjacent Neuproz groupings?

And finally, isotopic relaxing is simply the release of a neutron to form an isotope of the same element, which usually occurs in doubles as a result of Earth's bimodal magnetosphere. But more importantly, isotopic relaxing is DEPENDENT on the electronics specific to the Stereo Isotope. All of which means, the electronics of a Stereo Isotope determines if the isotope undergoes isotope relaxing OR a major/minor decay. Also, since Stereo Isotope electronics is based upon the environment, that means chemical bonding, magnetosphere, and temperature are also factors in the decay of Stereo Isotopes.

In conclusion, the Isotopic Degradation Pathway is complicated by Stereo Isotopes, which contain different atomic orbital arrangements that impart different levels of plasma to different regions of the atomic nucleus such that different degradation events occur, i.e. expansion of atomic particles and their expulsion or decay via the release of negative Neuprotrons. All of which, is the reason why the Isotopic Degradation Pathway isn't as clear cut as most scientists would like it to be. Butt, based upon the 4-particle difference between Neuproz groups, the abundances of the EVEN isotopes, and the placement of ODD elements, I was able to find a general trend and then extrapolate it to explain a majority of the periodic table. And finally, the association of extra neutrons about specific Neuproz groups probably facilitates similar Layer/groove electronics in the extra neutrons.

Chapter 6: Proximity Decays

You might not know this, but I've recently taken to making my bed in the morning. Granted, I only fluff the sheet over my four pillows, but that is beside the point. The point is, I often forget that I have a ceiling fan because it's really quite. And today, as I fluffed my sheet, it caught the ceiling fan, which extruded a massive ca-chunk sound. Thankfully, I removed the sheet from the blades of the ceiling fan and the ceiling fan returned to its normal rotation. Unthankfully, the ceiling fan developed a distinct and annoying 'thump' to its rhythmic rotation. Therefore, I turned the ceiling fan off, let it completely stop, and turned it back on again. Unfortunately, the annoying rhythmic 'thumping' persisted and the ceiling fan eventually stopped working. Fortunately, this is a great metaphor of proximity decays. But before I get to all that, let me take a moment and rendition your mind with a closely related idea.

Previously, I postulated a method to the forging of heavier elements, from lighter elements, in deep-dark-COLD positive space, which is based upon energy conservation. Or in electronic terms, protons stabilize lighter elements, but naked protons explode in deep-dark-COLD positive space, which results in lighter elements undergoing fission. As a result of lighter elements undergoing fission and spewing protons and neutrons in every direction, heavier elements can be forged via the electronic incorporation of inclusionary atomic particles over millions of years. All of which is a function the POSITIVE environment in deep-dark-COLD positive space that allows for the enhanced velocity of the atomic particles being spewed out in every direction such that they can from inclusionary elements. Unfortunately, the concept of negative Neuprotron layers really creates a quagmire to the formation of heavier elements when VIEWED definitely. Therefore, let's take a moment and expand our horizons such that the quagmire can be viewed as an integral part of the Universe's energy conservation.

To begin this journey of enlightenment, let's review the postulate that results in tons of cold rocks in deep-dark-COLD positive space: Not all stars go boom when they die, i.e. most stars just shrink and begin to forge heavier elements until they cool off, crack, and fall apart. All of which, has impregnated deep-dark-COLD positive space with the "soil" that humanity has grown on, i.e. Earth. Unfortunately, before I can go any further, I need to talk about the spectrum of protons as it relates to temperature.

Mass is a function of positivity expanding in this negativity negative branch of the universe. And as such, the electronics of protons are different within negative stars as compared to deep-dark-COLD positive space. Or in simpler terms, the negative Neuprotron grooves in colder protons/neutrons are probably different in protons that exist in negative stars. As for the reason why? Well, protons still have that "new-matter-smell" when they are surrounded by all the negative plasma in young-HOT stars. (FYI, based upon the negative environment, it is probably no surprise that the only matter forged in young-HOT stars is via negative Neuprotron Layer 1, i.e. deuterium and helium. All of which makes me wonder if a decrease in temperature opens up the subsequent negative Neuprotron Layers, i.e. 2-7.) In any event, the NEW matter in young stars is PERFECT for the formation of heavier elements in OLD stars. But, USED matter, i.e. lighter elements formed via the degradation of heavier elements, is still an integral part of energy conservation in the Universe. (Wait, have I used that sentence before?)

Let's suppose that some NEW heavier elements, from a recently departed star, find their way into a young star's magnetosphere. After millions of years of continual degradation, the second star dies and releases more NEW matter, which is mixed in with a bunch of used matter, i.e. lighter elements. Now, as the lighter elements degrade in deep-dark-COLD positive space, the USED protons/neutrons that existed about the NEW heavier elements maybe TOO old to aid in the formation of NEW Neuproz groups in even heavier elements. For example, oxygen degrades in deep-dark-COLD positive space and has the following Neuproz description: 2[N]2[3N]. Or in negative Neuprotron terms, oxygen has 12 atomic particles that are exchanging negative Neuprotrons via Layer 2 as well as 4 atomic particles that are exchanging negative Neuprotrons via Layer 1. All of which, brings to mind a certain level of quandrific intellectualism, which should not be confused with quadratic intellectualism.

First and foremost, can USED atomic particles ONLY forge Neuproz groupings about their USED layers? Or in simpler terms, once an atomic particle becomes part of a smaller Neuproz grouping, i.e. exchanging a LOWER energy negative Neuprotron, is it FOREVER locked into ONLY sharing negative Neuprotrons via the LESS energetic Layer? Or in the simplest terms, can the atomic particles from oxygen, which were sharing negative Neuprotrons via Layers 1 & 2, become inclusionary particles that eventually share negative Neuprotrons via Layers 6 & 7 in heavier elements? In any event, the short answer is maybe and the long answer is complex.

Depending on the amount of degradation within neutrons, it is possible that USED neutrons can participate in the exchange of negative Neuprotrons via more energetic Layers. As for USED protons, they are useless broken pieces of matter than can only be USED to warm the cockles of matter in deep-dark-COLD positive space. Unfortunately, all of this talk of 'USED atomic particle' incorporation as a method to the formation of heavier elements in deep-dark-COLD positive space is a kin to target shooting in the dark. Or in weirder terms, the atomic fission of lighter elements in deep-dark-COLD positive space spews USED atomic particles **randomly**, which does a **horrible** job at arranging USED atomic particles perfectly to _forge_ heavier elements. Butt, don't worry. Randomly placed USED atomic particles allow for the conservation of energy, albeit oddly, i.e. this odd path picks up near the end of this chapter.

The _absence_ of negative thermal energetic quanta is directly correlated to the **velocity** of protons and neutrons, which are released when lighter elements going BOOM in deep-dark-COLD positive space. And since the **velocity** of protons and neutrons is determined by the AMOUNT of peripheral negative energy, it is no surprise that atomic particles released in _negative_ environments move SLOWER, which determines the depth of insertion into adjacent elements, i.e. the atomic particles do **not** have enough velocity to scoot past all of the element's electrons. Therefore, semi-inclusionary elements are typically formed within negative environments, i.e. Star magnetospheres.

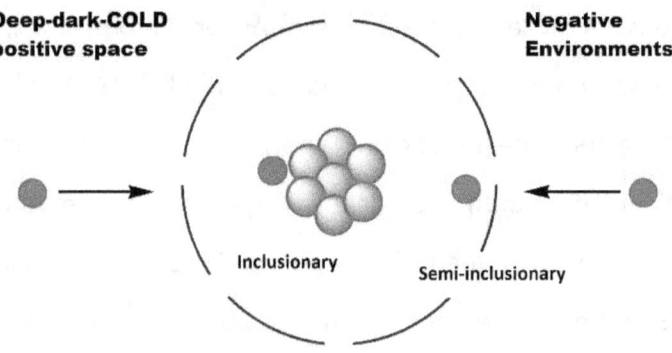

Figure 21: Semi-inclusion

As you can surmise from this figure, the presence of negative energy has slowed down the proton such that it only forms a semi-inclusionary element. Unfortunately, since protons are positive and electrons are negative, semi-inclusionary protons really fuck-up the negative atomic orbitals. And when the negative atomic orbitals are unable to adequately sate the positive charge within the atomic nucleus, neutrons are squirted out, i.e. isotopic relaxing. Or in simpler terms, semi-inclusionary protons create rhythmic atomic orbital thumps, which causes isotopic relaxing...among other things. (Foreboding!)

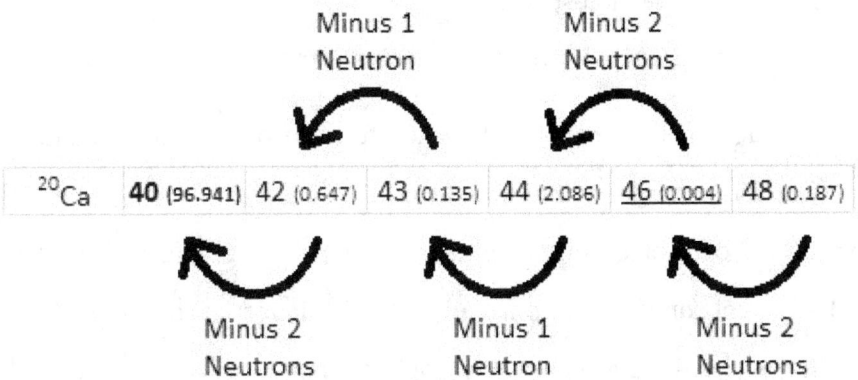

Figure 22: Isotopic Relaxing

As you can surmised, calcium 40 is the most relaxed calcium isotope, which is a function of atomic nuclei structure and environment. Also, you can see that the isotopic relaxing of calcium releases EIGHT protons, which are released as neutrons that quickly degrade into protons. All of which, DRASTICALLY influences the elements surrounding calcium to isotopically relax. Or in simpler terms, isotopes

41

relax and fart out protons based upon the environment, which creates semi-inclusionary elements that eventually isotopically relax and fart out more protons based upon their oddly-thumping atomic orbitals. Or in the simplest terms, relaxing isotopes cause adjacent atoms to isotopically relax. And while we're on the topic of isotopic relaxing, I might as well mention some other factors that facilitate isotopic relaxing.

1. Vibration
2. Pressure
3. Bimodal magnetospheres
4. Atomic orbital distortion by negative energy.

Pretty much, anything that fucks with atomic orbitals being able to sate the positive charge within the atomic nucleus, as it relates to the expansion of matter in the presence of negative energy, results in isotopic relaxing. In particular, let's investigate how Earth's magnetosphere plays a part in isotopic relaxing.

As a result of Earth's bimodal magnetosphere, elements are more apt to release two neutrons to maintain symmetry. All of which means, **one** semi-inclusionary proton may result in the release of two or more neutrons, which degrade into protons. But, there are exceptions to the rule. For example, silicon does not adhere to the trend of symmetric isotopic abundance, which is caused by Earth's bimodal magnetosphere.

Element	Isotope Weight (Abundance)		
^{14}Si	28 (92.23)	29 (4.67)	30 (3.10)
^{8}O	16 (99.762)	17 (0.038)	18 (0.2)

Table 4: Exceptions to the rule[1]

As you can see in this table, silicon-29 is more abundant than silicon-30. Quite simply, silicon exists in nature as a tetra-oxide polymer, i.e. sand. And as a result of this, silicon is electronically protected from Earth's bimodal magnetosphere as well as protected from semi-inclusionary protons because oxygen is more electronegative than silicon. Or in simpler terms, oxygen's negativity sucks up most of the protons/neutrons that are being released by relaxing isotopes, which is one reason why oxygen is ALMOST isotopically pure. Other reason why oxygen is ALMOST isotopically pure is because it has shaken-particle-syndrome (SPS). Or in other words, TONS of heavier elements form oxides, which means when the chemical bond vibrates, the lighter oxygen gets shaken around. And when oxygen gets shaken around a lot, it weakens the weak columbic nuclear forces, which causes the expulsion of neutrons, i.e. isotopic relaxing.

Another interesting consequence of oxygen being electronegative and sucking up protons being released by isotopic relaxing is: Earth has a lot of nitrogen. Or in simpler terms, the continual formation of semi-inclusionary oxygen atoms has modified the half-life of oxygen, i.e. forged nitrogen in a timely manner...whatever that means.

Element	Isotope Weight (Abundance)		
^8O	16 (99.762)	17 (0.038)	18 (0.2)
^7N	14 (99.634)	15 (0.366)	

Table 5: Birth of a Base[1]

As a result of the continual bombardment of oxygen with semi-inclusionary protons, ^{16}O decays into ^{15}N, which isotopically relaxes into ^{14}N. And for those of you who don't know this, the silicon-nitrogen bond is weaker than the silicon-oxygen bond, which means silicon will typically release nitrogen and then bind to a different oxygen, i.e. a method to Earth's weird abundance of atmospheric nitrogen gas, which will be discussed in Chapter 8.

All of which, brings us back to the oddity at the beginning of this chapter. Quite simply, some USED atomic particles may or may not intercalate into heavier elements. If they do intercalate, then they're probably responsible for forging smaller Neuproz groups and/or becoming extra neutrons that have weak columbic nuclear forces with their neighbors. And if they don't intercalate, then they simply exist as inclusionary particles, which are held in place by very weak columbic nuclear forces. Either way, their presence modulates atomic orbital placement and results in: Magnetically active elements and/or pseudo-isotopes that have an enhanced rate of decay, i.e. atomic particle ejection, which results in semi-inclusionary elements that are probably magnetically active. All of which results in energy conservation.

For a moment, let's imagine a large rock that has been basking in deep-dark-COLD positive space for millions of years. Unfortunately, as a result of this and the fact that atomic particles only have a SPECIFIC number of negative Neuprotrons, a lot of the elements in this rock have inclusionary protons and neutrons, i.e. they are pseudo-isotopes. As a result of the extra inclusionary particles in the pseudo-isotopes, they having distorted atomic orbitals that are MORE responsive to negative energy, i.e. allow negative energy to access the atomic nucleus. And when more negative energy innervates the pseudo-isotope's atomic nucleus, it takes LESS energy to cause pseudo-isotopes to decay, release atomic particles, and forge adjacent semi-inclusionary elements that are magnetically active, which facilitates the movement of cold rocks towards warm stars. (BTW, I'll talk about this in more detail in the next chapter.) Also, pseudo-isotopes are quicker to decay and release positive protons, which are attracted to the negatively diffusive stars. And finally, USED atomic particles can decay to yield a little negativity to the rocks in deep-dark-COLD positive space. In any event, USED atomic particles are a factor in energy conservation regardless of their ability to forge more energetic Neuproz groupings, i.e. Neuproz groupings that are exchanging negative Neuprotrons via Layers 6 & 7.

In conclusion, pseudo-isotopes have an enhanced sensitivity to negative energy because the inclusionary atomic particles distort the atomic orbitals, which allows more negative energy to reach the atomic nucleus. All of which means, miniscule amounts of negativity can cause pseudo-isotopes to become magnetically active and/or degrade via the release of atomic particles, which can cause adjacent atoms to become semi-inclusionary elements. As for the benefit of semi-inclusionary elements? The rhythmic disturbance in atomic orbital placement results in isotopic relaxing, which releases MORE atomic particles to form OTHER semi-inclusionary elements. Also, the increase in positive protons causes the rock to be attracted to negatively diffusive stars. And finally, the only thing that protects elements from being involved in proximity decays is by the presence of electronegative oxygen, which creates a protective electronic environment. Unfortunately, oxygen usually goes boom in deep-dark-COLD positive space so cold rocks are more responsive to the degradation of pseudo-isotopes. (Is it weird how everything ties together? In any event, the next chapter will tie pseudo-isotopes to organized planetary cores that facilitate the formation of bimodal magnetospheres via semi-inclusionary elements, which helps in isotopic relaxing, i.e. FUN ON A BUN!)

Chapter 7: Planets

Until recently, magnetism was thought to be a magical force that oddly perforated the universe. But, since we exist in a negative branch of the universe, magnetospheres are created by negative magnetic energetic quanta. And as such, the presence of magnetism influences elemental half-lives and type of isotopes in planets. All of which, paints a very intricate picture of atmospheres within our solar system, which will be covered in the next chapter. In this chapter, let us muse about the origins of planetary bimodal magnetospheres.

With the slightest amount of negativity, pseudo-isotopes become magnetically active due to abnormal atomic orbitals. Also, negativity causes pseudo-isotopes to degrade and release protons, which create magnetically active proximal semi-inclusionary isotopes. And as a result of the attraction of opposite magnetic energetic quanta, there is an increase in negative plasma between the cold rock and star. All of which, allows for atomic orbital modulation to push the rock towards the star. In any event, negativity causes cold positive rocks to push towards negative stars, which warms the rock up to a liquidity state. Unfortunately, before get into the trajectory of non-planets, I need address the concept of molten gravity.

Everybody knows that Earth has a dense core. And I assume that most people believe that gravity PULLED all the heavier elements towards the Earth's core to make it dense. I, on the other hand, believe something different. I believe that the density of Earth's core was forged via reverse size exclusion, i.e. the smaller elements move faster away from the Earth's core. As for the method of this reverse size exclusion densification, please scrutinize the figure below.

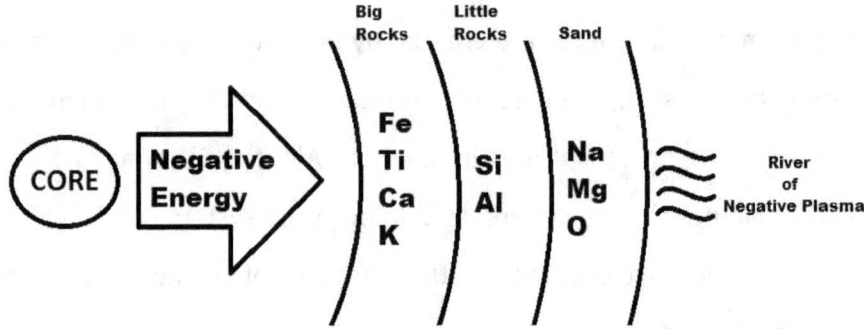

Figure 23: Molten Gravity

As you can see in this figure, the continual diffusion of negative plasma away from the core results in the pronounced movement of smaller elements, i.e. the accumulation of heavier elements about the core. Or in simpler terms, the core produces a river of negative plasma, big rocks barely move, little rocks move a little, and sand EASILY follows the current of the negative plasma. Or in scientific terms, the degradation of pseudo-isotopes results in semi-inclusionary elements that cause electrons to degrade and release negative energy, which DIFFUSES away from the core and percolates the smaller elements to the surface, thus leaving the heavier elements about the core. (BTW, that includes percolation of SMALL positive protons, which aids in the attraction of the rock towards the negative star and SUCKS the negative out of the rock's core.) In any event, molten gravity is simply lighter elements, i.e. sand, following the river of negative plasma. All of which, brings us back to the trajectory of non-planets.

For those of you who don't know this, planets have <u>circular</u> trajectories and NON-planets have <u>oval</u> trajectories about negatively diffusive stars. As for the reason why this is important? Well, NON-planets don't have magnetospheres to stabilize circle trajectories. As for the METHOD by which non-planets develop bimodal magnetospheres? Well, as liquid cores become denser, which happens periodically because their comet-esk oval trajectory results in repetitive heating and cooling, the elements align synergistically as a result of REPETATIVE crystallization. Or in scientific terms, God knows crystallization is an art, not a science. In any event, repetitive crystallization results in a synergistically aligned core, which develops a magnetosphere and places the rock in a spherical orbit about the negatively diffusive Star. All of which, brings us to an intellectual brawl: Some planetary cores are COLDER than their magma layers. (I'm not sure who I'm brawling with most of the time, but that is beside the point.) The point is, based upon the trajectory of objects without magnetospheres, i.e. their repetitive cycle of heating and cooling, the concept of Entropy, the exposure of the outside of the rock to the negatively diffusive Star, and the fact that heavier elements have immense thermal capacities, some planetary cores are COOLER than their magma layers. Actually, let me try and explain that a little slooooowwwwweeeeerrrr.

Heavier elements come from deep-dark-COLD positive space. Also, heavier elements have very HIGH thermal capacities, which may sound counter to my logic unless you realize that it takes a TON of energy to FORCE negative thermal energetic quanta into a horde of heavier elements. Also, heat diffuses quicker from crystalline substances in comparison to amorphous substances, i.e. magma. In any event, via this logic, some planetary cores are COLDER than the magma layers because the negative energy in the magma layer, which is the result of the Sun, quickly diffuses into deep-dark-COLD positive space. (BTW, magma layers remain immensely hot for millions of years because the presence of negative energy facilitates elemental fission, which can be a cascading process.) Unfortunately, I have to stop here and reiterate the errors of past theories such that you don't blow your cork and stop reading.

Old theories of gravity believed that energy was being "beamed" from the Sun into Earth's core, which made Earth's core super-HOT. Or maybe the Sun's gravity was grinding Earth's core together to cause friction and heat? Whatever the case maybe with regards to the old theories, I've postulated that gravity has many components. One of them being the attraction of **opposite** magnetic energetic quanta, which are released when electrons are stimulated to decay via the collision with negative thermal energetic quanta. All of which, returns us to the point at hand: How does a dense COLD synergistically aligned core of heavier elements create a magnetosphere? Wait, is that an explanatory figure I see? Damn, how did that get there?

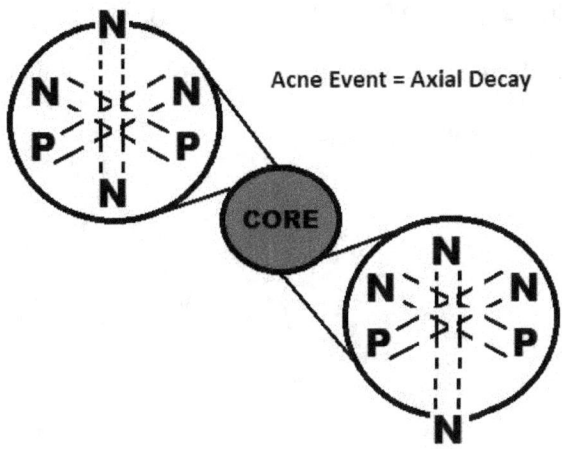

Figure 24: Magnetic Initiation

As you can arrange from this figure, the blue CORE represents a collection of COLD heavier elements that are aligned synergistically such that opposite neutrons are being released at opposite sides of the core. But more importantly, it is the extreme heat in the magma layer which is causing the decay of the elements in the colder blue CORE, which is the initiation step to the formation of a bimodal magnetosphere. Please read the next paragraph to understand why.

Some elements are magnetic and some elements are not magnetic. And since every element has a unique arrangement of atomic orbitals, it is safe to say that magnetism is a function of atomic orbital placement. More specifically, I've postulated that certain elements have 'wobbling' atomic orbitals that cause electrons to repeatedly crash into negative thermal energetic quanta, which causes them to degrade and release magnetic energetic quanta. Therefore, with all that in mind, it should be easy to imagine how a semi-inclusionary proton in a non-magnetic element might cause the atomic orbitals to WOBBLE to make it magnetically active.

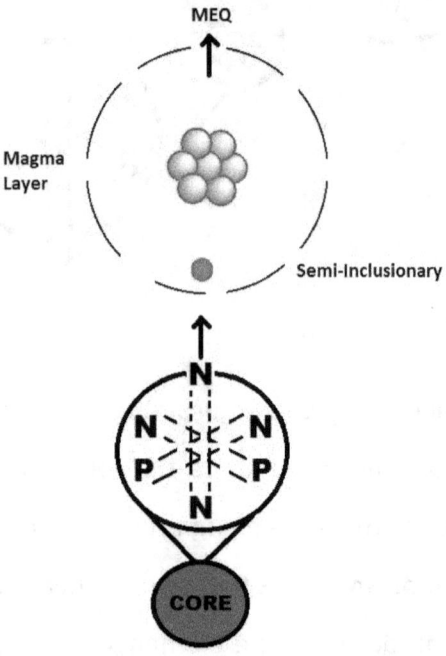

Figure 25: Magnetic Amplification

As you can see in this figure, the release of a neutron creates a semi-inclusionary element within the magma layer, which BECOMES magnetically active via atomic orbital distortion, i.e. the AMPLIFICATION step to planetary magnetospheres. Or in simpler terms, semi-inclusionary protons cause elements to release one or

MORE neutrons, which can form more semi-inclusionary elements and more magnetically active elements within the magma layer. As for the directional specificity that creates Earth's bimodal magnetosphere? Well, that gets a little complicated.

I think we can all agree that Earth's core is dense. Yes? Well, one consequence of dense elements is that they tend to align electronically, i.e. adjust the placement of their atomic orbitals such that they are less intrusive to their neighbors. (If only we were more like atoms.) And as a result atomic orbitals being a function of atomic nuclei structure, when an atom on the outer layer of Earth's dense blue cold core releases a neutron, the atomic orbital structure changes.

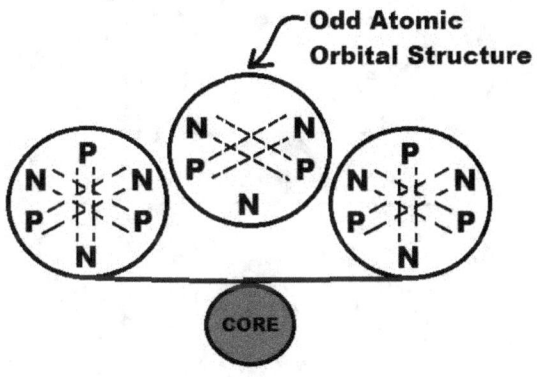

Figure 26: Directional Synergy

As you can see in this figure, the middle atom has lost a proton, but it is still being held in place by its neighbors' atomic orbitals. As a result of this, the element is forced to have an odd atomic orbital arrangement, which results in directional magnetic activity via atomic orbital wobbling. But, even as this atom's neighbors lose their protons, the

second layer of the dense cold core imparts an odd atomic orbital arrangement to the <u>first</u> layer of the dense cold core...until such time the elements in the second layer of the dense cold core decay and push the first layer of elements into the magma layer. All of which means, the synergistic arrangement of the elements within the dense cold core result in opposite spin electrons degrading at either ends of the dense core.

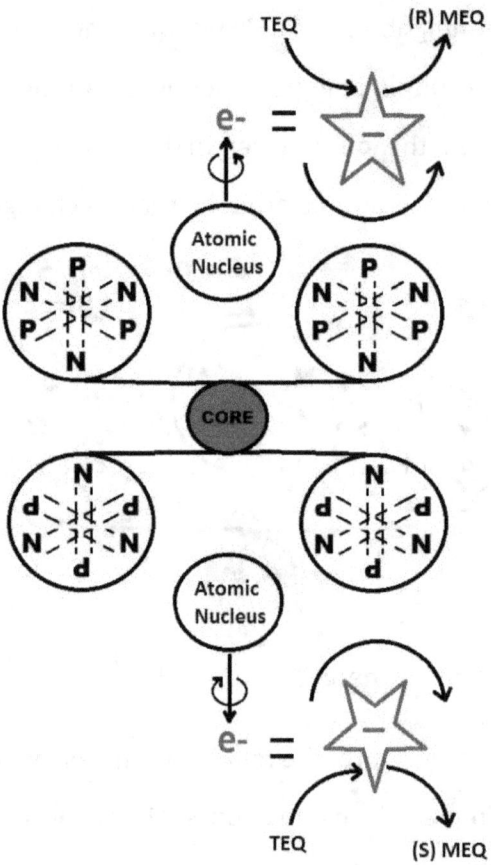

Figure 27: Magnetic Alignment

As you can see in this figure, the synergistic alignment of the elements in Earth's core results in the exposure of OPPOSITE spin electrons at either ends of the dense cold core, which provides magnetic **alignment** to the magnetically active semi-inclusionary elements within the magma to forge Earth's bimodal magnetosphere. All of which, leads to the final step to planetary magnetism: The ability to switch.

In as much as much as Earth's dense core is a collection of synergistically aligned heavier elements, there are varying layers of electronic alignment. Or in simpler terms, look down.

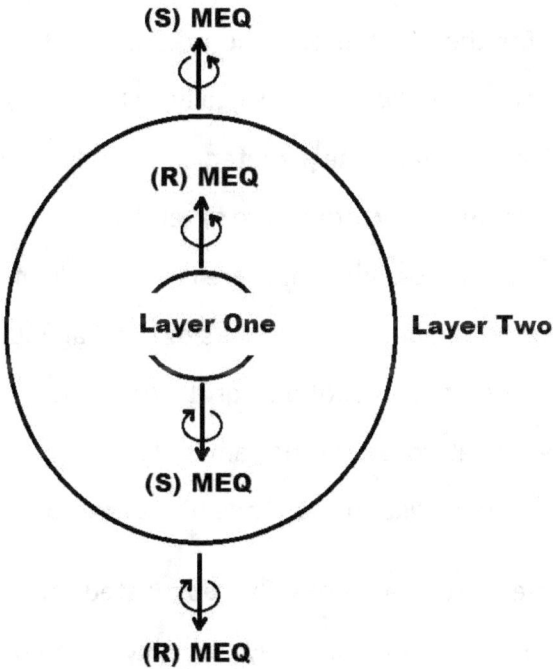

Figure 28: Core Layers

As you can clinch from this figure, the slow degradation and diffusion of the dense cold core layers into the hot magma will eventually result in the **inversion** of the magnetic field. Or in simpler terms, initiation and amplification events will be exactly the same, but the alignment will be different. (BTW, as the dense cold core shrinks and the magma amplification layer increases, the planet will form a stronger magnetosphere...until a point. I'll talk about that in the next chapter.)

With all these postulates on the table, I would like to add some realism to these picture perfect postulates. For example, the alignment of the dense cold core's heavier elements is not only dependent on the heavier elements therein, but also the pressure and time the heavier elements are given to align synergistically. Or in simpler terms, it is theoretically feasible that some planetary cores only a have a few layers of synergistic alignment and others are synergistically aligned the whole way through. Or, it is possible that there is a few million years between synergistic layers, i.e. a few million years of NO magnetosphere, which would cause the planet to cool and drift further away from the Sun. Unfortunately, not all cores are the same. Butt, before I get to all that in the next chapter, I'd like to share one more absurd postulate.

Over the course of several books, I've postulated how the universe is created via the expansion and decay of highly ordered energy. All of which, began with God's pet singularity developing acne, our negative branch of the universe being the result of an odd zit, and our galaxy being charged with negatively diffusive stars. But, upon further reflection with regards to MOST old stars forging heavier elements instead of exploding, i.e. energy conservation, it is plausible that a majority of planets are simply the hearts of dead stars...or at least really large chunks of dead star hearts. As for the rationale behind this postulate? Well, it is pretty straight forward. First and foremost, dying stars form heavier elements and they have lots of negativity energy to produce Molten Gravity, which forges dense cores. Next, our solar system is really close to the edge of one of Milky Way's swirls. And finally, the continual crystallization of planetary cores to facilitate the formation of bimodal magnetospheres is SEVERELY complicated by the cohesion of smaller asteroids. Or in simpler terms, non-magnetic COLD rocks do NOT have circular trajectories such that the rocks can warm up and congeal and form large planetary bodies. As a result of this, the ONLY way to congeal a mass from cold rocks is via the formation of a magnetically active warm planetoid and then bombarding it with cold rocks, which will SEVERELY fuck with the **order** of the planetary core that is creating the bimodal magnetosphere. In any event, like I said, it is an absurd postulate. But, in terms of our swirl-ish Milky Way galaxy sweeping up the hearts of dead stars to form planets, it DOES seem plausible. Actually, the true dissonance to this postulate is based around the SIZE of cold rocks in comparison to dead star hearts, i.e. dead star hearts are just big cold rocks.

In conclusion, pseudo-isotopes are more sensitive to negative energy based upon their atomic orbitals, which causes them to initiate a gravitational response to negatively diffusive stars via release of positive protons and the production of magnetically active semi-inclusionary elements. As a result of these gravitational components, the rock takes a comet-esk trajectory about the negatively diffusive star, which allows for the synergistic alignment of the core's heavier elements via repetitive crystallization. Also, the densification of the rock is caused by Molten Gravity, which is just reverse size exclusion based upon the diffusion of negative energy. Eventually, the magnetosphere is created and amplified by the semi-inclusionary elements in the magma layer, which align to the directionality of opposite electrons in the cold dense core.

Chapter 8: Magnetospheres

With the crystallization of bipolar magnetospheres in your noggin, it is time to ask a difficult question: Is gravity directly proportional to magnetosphere strength? Or in more direct terms, gravity is **NOT** directly proportional to magnetic strength, i.e. magnetic strength is a component of gravity. Or in the simplest terms, look down.

	Mercury	Venus	Earth	Mars	Jupiter	Saturn	Uranus	Neptune	Pluto
Distance (km)	56848000	107712000	149600000	227392000	777920000	1421200000	2872320000	4488000000	5909200000
Radius (km)	2433.722	6046.079	6371	3395.743	71291.49	60269.66	25356.58	24273.51	1121.296
Mass (kg)	3.28625E+23	4.86963E+24	5.975E+24	6.39325E+23	1.90005E+27	5.67625E+26	8.9625E+25	1.01575E+26	1.49375E+23
Density (g/cm^3)	5.43	5.25	5.52	3.93	1.33	0.71	1.24	1.67	2
Pressure (bars)	2E-15	90	1	0.006	100000	100000	100000	100000	2E-14
Temperture (K)	400	730	290	225	124	95	58	59	50
Moons	0	0	1	2	16	17	15	8	1
Atmosphere	31.7% K	96.4% CO_2	78% N_2	95.3% CO_2	81% H_2	93% H_2	82% H_2	84% H_2	CH_4
	24.9% Na	3.4% N_2	20.9% O_2	2.6% N_2	17% He	5% He	14% He	12% He	
Gravity (m/sec^2)	3.7	8.87	9.78	3.72	22.88	9.05	7.77	11	0.4

Table 6: Planetary Data[1-3]

As you can see if this figure, if all the numbers don't make your mind go numb, Saturn is more massive than Neptune, but Neptune supposedly has more gravity than Saturn, i.e. 11 and 9.05 respectively. Or simpler terms, magnetospheres have the ABILITY to contain elements, but if there are no elements to collect, then magnetism has less of an impact on gravity. For example, let's imagine we could create a hallway with a FIXED magnetosphere in it.   Now, if the hallway contains NO atmosphere, then there is NO resistance to you running through it...except for your suffocation. But, if we fill the hallway with Krypton, i.e. a really heavy noble gas, then it will take LARGE amount of force to push through the heavy Krypton...especially if you're Superman. Therefore, it takes more energy to push through helium because it is heavier than hydrogen, which could be a factor as to why Neptune, 12% He, has more gravity than Saturn, 5% He. In any event, let's skip to point of this chapter: I postulate that Saturn has the strongest magnetosphere in our solar system. Unfortunately, there is a caveat to this postulate: Jupiter has the strongest magnetosphere CREATED by the decay of heavier elements in a COLD dense core and Saturn has the strongest magnetosphere created by the synergistic alignment of WARMER lighter magnetically active elements.

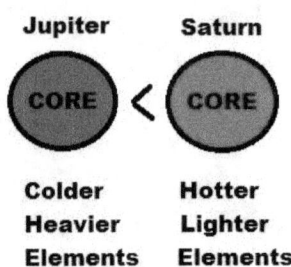

Figure 29: Magnetic Strength

As you can see in this figure, it is the age old debate of Republican verses Democrat. Wait, what I meant to say is: Saturn's warm red core creates a **stronger** magnetosphere than Jupiter's cold blue core. As for the basis of this postulate, it is pretty straight forward. Cold **dense** cores are initiating the DIRECTIONALLY of the bimodal magnetosphere via continual degradation, but the DENSITY of the core destroys some of the magnetism released in the magma layer. The warm core on the other hand, does **NOT** have a **dense** core to destroy any magnetism, which means the synergistic alignment of the warmer elements is stronger. All of which, is BASED upon density.

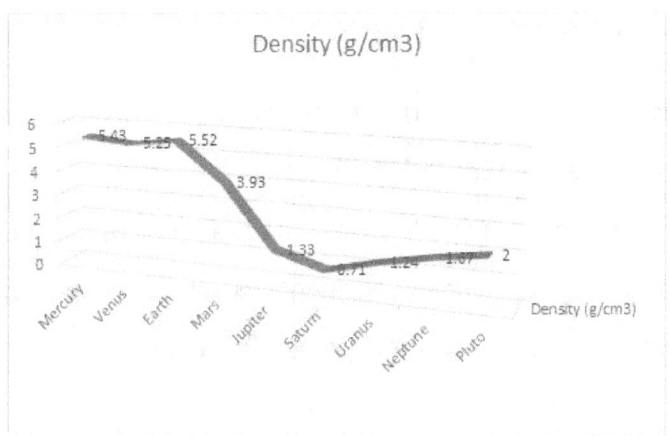

Figure 30: Planetary Density

As you can see in this figure, Saturn's density is oddly lower than Jupiter and Uranus. Butt, Saturn has the most moons, some really cool rings, and the highest level of atmospheric hydrogen gas in the solar system. And just in case the importance of these facts escape you, let's review their theological importance.

1. Moons don't hang around without a reason.
2. It takes a lot of energy to align cold small rocks about the mid-plane of a planet.
3. Hydrogen gas is normally magnetically NON-responsive, i.e. it takes a very strong magnetosphere to contain it.

All of which, is the basis of my postulate that Saturn has the strongest magnetosphere. (FYI, if you were to plot the temperature and density of the planets from Saturn to Pluto, you'll see that as the temperature drops, so does the ability of the planet to have an atmosphere, i.e. Pluto doesn't have much of an atmosphere.) In any event, here are the major factors to magnetosphere strength:

1. Distance from a negatively diffusive star.
2. The arrangement of the core, i.e. dense decaying OR synergistically aligned. (BTW, the magma amplification layer is important to dense decaying cores and temperature is important to synergistically aligned cores.)
3. The elements within the atmosphere.
4. Density, thickness, and composition of the crust.

And while we're on the topic of atmospheres, I need to make an odd postulate: Jupiter's black spot storm, which has been raging for thousands of years, is simply a collection of NON-greasy elements within Jupiter's greasy atmosphere. Or in simpler terms, the hydrogen gas in Jupiter's atmosphere is the grease and the elements occluded from this greasy atmosphere, i.e. the black dot storm, are simply a collection NON-greasy elements. Or in the simplest terms, Jupiter probably had a volcanic eruption that spewed a thick dense cloud of NON-greasy elements into the upper atmosphere, which have not precipitated to the ground because of their continual electronic reaction with all the hydrogen gas. (BTW, if Saturn had a NON-greasy spot storm, it probably precipitated because its atmosphere is like 93% hydrogen gas.) As for the reason for all the hydrogen gas in 'older' planets? Well, stronger magnetospheres result in isotopic relaxing, i.e. the release of neutrons, which decay into protons to form hydrogen gas. Also, older planets have MORE lighter elements, which can be stimulated to isotopically relax by being shaken by heavier elements.

In conclusion, magnetism is simply a component of gravity. For example, heavier elements in the atmosphere means it will take more force to push through those heavier elements. As for magnetic strength, it generally varies as a result of: Type of core, distance from the Sun, temperature, and the crust of the planetoid. And finally, it is worth mentioning that different atomic orbitals will degrade different types of magnetic energetic quanta, which means different atmospheric compositions will play a part in magnetosphere strength. Or in simpler terms, if Earth had more hydrogen gas in its atmosphere, it would have a stronger magnetosphere because hydrogen gas is gravitationally NON-responsive and will degrade LESS negative magnetic energetic quanta.

Chapter 9: Compartmentalization

A long time ago in a galaxy far far away, I put forth the following premise: There is a method to God's madness. Unfortunately, the madness that surrounds humanity's cognition of time has got me into a little bit of trouble. (Most of us are lucky if we live long enough to see Uranus make one revolution around the Sun.) Butt, I refuse to learn. Therefore, I postulate that God experiences DIFFERENT time because he/she/it is NOT constrained by the laws of energy conservation, i.e. God creates energy. (I believe in God, but nobody believes that.) All of which, allows me to believe in human evolution as a function of planetary evolution, which is a function of isotopic evolution as it relates to unique environments. Or in simpler terms, this chapter is about evolution.

The first question that needs to be addressed with regards to evolution is: Where does carbon come from? Well, by the Isotopic Degradation Pathway, isotopic relaxing resulted in Earth's magnetosphere, which amplified isotopic relaxing to produce lighter elements. Unfortunately, since carbon is instrumental to life as a means of data storage, carbon-14 degradation is not conducive to life and/or data storage. Therefore, how did Earth end up with the isotopic purities that are conducive to life, i.e. carbon-12 is 98.9% isotopically pure on Earth?[1]

If you go back and gander Table 6 in Chapter 8, you'll see that Venus's atmospheric composition is about 96% carbon dioxide. Therefore, if Earth evolved via a Venus-esk state, then the isotopic purity of oxygen is probably the result of the vibration of oxygen about carbon dioxide, which is more exposed to the planet's bimodal negative magnetosphere. (FYI, the vibration and association with other elements ALSO probably aided to oxygen's isotopic purity.) And if the vibration of carbon dioxide aided in the isotopic purification of oxygen, then more than likely it had a part in the isotopic purification of carbon. (Thank you carbon dioxide?)

Human Composition	Element	Isotope Weight (Abundance)					
61%	^{8}O	16 (99.762)	17 (0.038)	18 (0.2)			
23%	^{6}C	12 (98.90)	13 (1.10)	14 (0.0)			
10%	^{1}H	1 (99.985)	2 (0.015)				
2.60%	^{7}N	14 (99.634)	15 (0.366)				
1.40%	^{20}Ca	40 (96.941)	42 (0.647)	43 (0.135)	44 (2.086)	46 (0.004)	48 (0.187)
1.10%	^{15}P	31 (100.0)					
0.20%	^{16}S	32 (95.02)	33 (0.75)	34 (4.21)	36 (0.02)		
0.20%	^{19}K	39 (93.2581)	40 (0.0117)	41 (6.7302)			
0.14%	^{11}Na	23 (100.0)					
0.12%	^{17}Cl	35 (75.77)	37 (24.23)				
0.027%	^{12}Mg	24 (78.99)	25 (10.0)	26 (11.01)			
0.026%	^{14}Si	28 (92.23)	29 (4.67)	30 (3.10)			
0.006%	^{26}Fe	54 (5.8)	56 (91.72)	57 (2.1)	58 (0.28)		
0.0037%	^{9}F	19 (100.0)					
0.0033%	^{30}Zn	64 (48.6)	66 (27.9)	67 (4.1)	68 (18.8)	70 (0.6)	
0.0001%	^{29}Cu	63 (69.17)	65 (30.83)				

Table 7: Humanity[1]

As you can surmise from this table, 99.1% of the human body is composed by isotopes that are at least 98.9% pure.   But more importantly, the human body is 61% oxygen, which limits the deleterious effects of isotopic relaxing, i.e. the release of neutrons. Or in simpler terms, the warmth of the human body, sustained by the thermal capacity of water, as well as the electronegativity of oxygen, DECREASES the momentum of atomic particles being released by isotopic relaxing, which is kind of amazing when you think of it in those terms. All of which, leads us to the evolution of complex molecules that are the basis of life on Earth.

Another interesting thing about the evolution of Earth through a Venus-esk pathway is the decay of ONE oxygen about carbon dioxide such that it creates isocyanate, which gives us a viable method to the formation of complex heterocycles, i.e. DNA. (Again, thank you carbon dioxide?)

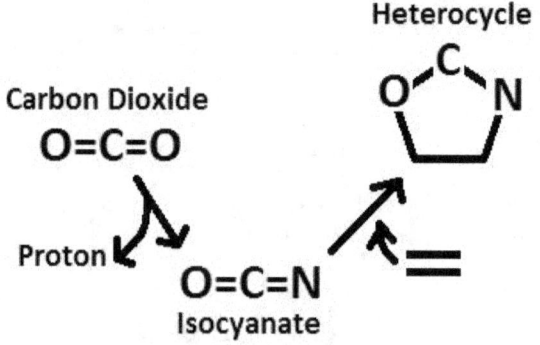

Figure 31: Isocyanate

As you can see in this figure, the decay of ONE of carbon dioxide's oxygens via the release of a proton results in the formation of ONE nitrogen, i.e. carbon dioxide degrades into isocyanate. And for those of you that didn't enjoy organic chemistry, isocyanates undergo a pericyclic reaction with double bonds to create a heterocycle. All of which bring us to the fundamental question: How did single stranded DNA and/or RNA occur? (FYI, complex heterocycles, i.e. DNA, probably occurred via a similar catalytic pathway.) The short answer is: Chelation. The long answer is: muy grande.

LIFE is dependent on catalysis and catalysis is usually dependent on transition elements. But, for anyone who has studied nanotechnology, you'll know that transition elements have a tendency forge repetitive arrangements, i.e. nano-technology. Therefore, all you have to do to understand the formation of single stranded DNA/RNA is realize the following: Heterocycles have nitrogen, nitrogen chelates transition metals, transition metals from nano-structures, and transition metals are catalytic. Or in simpler terms, look down.

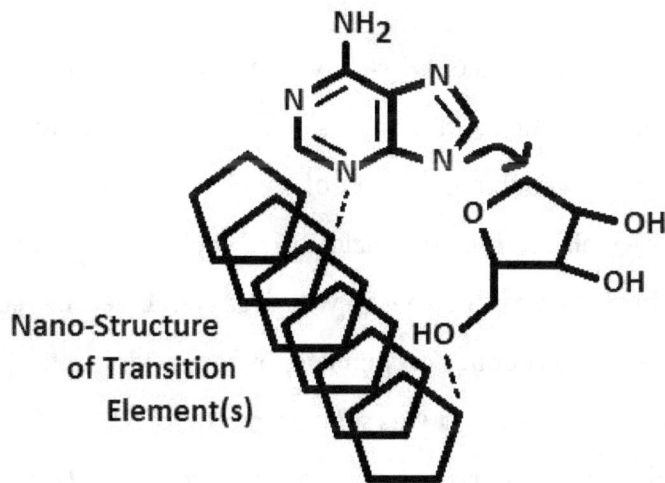

Figure 32: Proximity Rules

As you can see in this annoyingly simplistic figure, the constituents of DNA can be placed in the appropriate proximity such that DNA monomer is formed. And via this same logic, DNA monomers can be place in the appropriate proximity to forge single stranded DNA. Or in simpler terms, the nano-structure of transition metals provides a TEMPLATE and CATALYSIS for single stranded DNA. Or in the simplest terms, the coding of life resides in the characteristics and structure of HOW transition elements "polymerization" in different environments. All of which means, the energy released by isotopically relaxing elements on Earth provided more than enough substraights and mixing to form a plethora of different single stranded DNA, RNA, and proteins, which also contain chelating nitrogen components. All of which, flooded Earth with tons of different single stranded DNA/RNA subunits based upon the different nano-structures of transition elements.

In as much as life would be clunky if it contained large crystals of transition metals, the formation of single stranded DNA/RNA/proteins allowed for the variable chelation of smaller collections of transition metals, which allowed for the inclusion of these smaller subunits into greasy bubbles. As for the reason why greasy bubbles, i.e. cells, came about? Well, it's all about the chemistry of fatty acids, i.e. they naturally forge bubbles. Therefore, it was simply a matter of time before a grease bubble captured a catalysis, i.e. a transition metal chelated by single stranded DNA/RNA.

Now here is the beauty of the system: With a catalysis within a grease bubble, the catalysis was able to capture small molecules that diffused into the grease bubble to form larger molecules, which were NOT able to diffuse out of the grease bubble. All of which, resulted in the increase of material within the grease bubble and the eventual fissioning of the grease bubble into two grease bubbles. But, this is where things get tricky. The second grease bubble contained the molecules forged by the catalysis within the first grease bubble, which makes the second grease bubble completely unique from the first grease bubble.

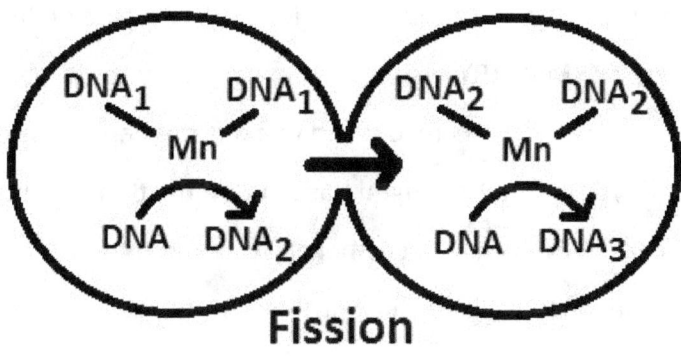

Figure 33: Different single stranded DNA/RNA

As you might not be able to surmise from this picture, DNA_{1-3} are single stranded DNA polymers. But, what you might be able to conclude from this picture is that the unique Manganese-DNA_1 catalysis is able to trap DNA monomers via catalyzing the formation of single strand DNA polymers, which cannot diffuse out of the grease bubble and causes the infusion of water into the grease bubble. As a result of this increase in volume, the grease bubble undergoes fission to produce a second grease bubble that contains another unique catalysis, Manganese-DNA_2, which catalyzes the formation of the DNA_3 polymer. Or in the simplest terms, the world's first mitosis event was caused by a little grease, some single stranded DNA/RNA, and a couple transition metals. Unfortunately, this next part is where non-scientists get hung-up.

After millions of years of grease bubbles containing more and more complex catalysis, simply because Earth was gurgling with energy, a self-replicating system evolved. Granted, mitosis is simply a FACT of increasing volume caused by catalysis, but something else happened: Internal complexity resulted in SYMMETRICAL fissioning. And somehow, this SYMMETRY resulted in better energy conservation. But, when the new "cells" were introduced to a different environment, their internal catalytic systems changed based upon the things that diffusing into the "cells". Eventually, cells began to associate because it made LIFE easier, i.e. one cell liked the shit that another cell released. Next, single cells started squirting out plasmids in an order to protect themselves, which resulted in some single cell organisms collecting extra data via supercoiling the excess plasmids, which result in chromosome-esk structures. And finally, a cell sucked up a mitochondria organism to provide 'unlimited' cellular energy and the rest is history. Granted, it is a crazy complex history of continual modification based upon DIFFERENT things diffusing into cells and the synergy of things SHARED between cells over millions and millions of years, but efficiency is usually rewarding in the absence of entities with sharper teeth. All of which, resulted in humans.

Each cell in the human body contains different proteins, which are the result of different genes being expressed. Unfortunately, the building blocks of these proteins, amino acids, do NOT diffuse across double layered cellular membranes. All of which means, each cell has a different AFFINITY for amino acids based upon the proteins sucking up the amino acids. Or in simpler terms, look down.

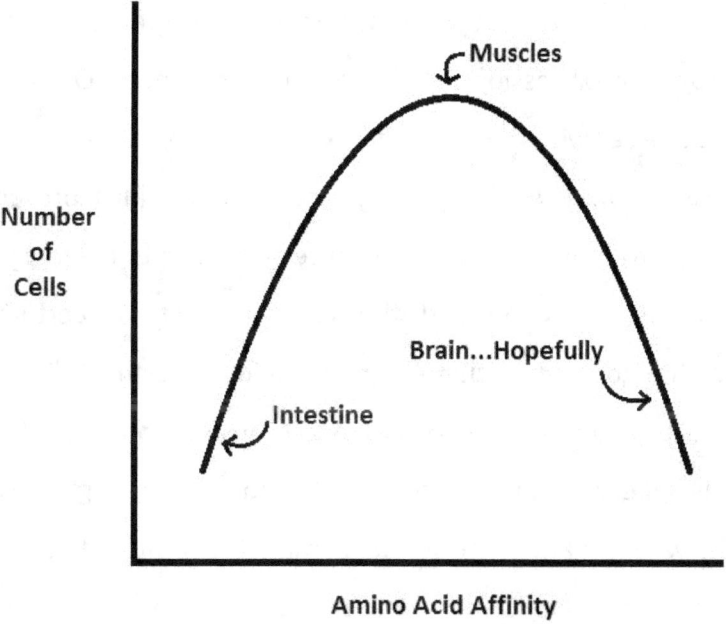

Figure 34: Amino Acid Affinity

As you can see in this figure, which is hypothetically based upon my unique body, each cell has a different affinity for amino acids.

Therefore, when the body is NOT getting enough amino acids, each cell is FIGHTING for the amino acids it needs to survive and thrive. And as a result of each person being unique, each person has a unique CELL that is NOT getting the amino acids it needs. All of which, can result in the cancer or tumors.

For example, one person might have a cell with the weakest affinity for certain amino acids within the brain, which could cause the cells to produce and secret non-functional proteins, i.e. Parkinsons and Alzheimers. Or, one person might have a cell with the weakest affinity for a certain amino acids within the testicles or breasts, which could change protein expression and the formation of cancer. Or, one person might have a cell with the weakest affinity for certain amino acids within heart tissue, which could cause abnormal extracellular proteins with weaker 'gluing' power such that it causes the heart to enlarge. Or, one person might have a cell with the weakest affinity for certain amino acids within connective tissue, which could cause the cells to secret abnormal proteins and the body to target these abnormal proteins with antibodies, i.e. rheumatoid arthritis. Or, one person might have a cell with the weakest affinity for certain amino acids within the intestines, which could cause abnormal extracellular proteins with weaker 'gluing' power, i.e. an increase in the exposure of the immune system to roughage in your intestines such that your body produces anti-bodies towards this harmless roughage, i.e. Celiac disease. Or, the possibilities could go on and on and on. (BTW, in as much as proteins determine the efficacy of DNA replication, protein errors can cause DNA errors, which can also cause cancer.)

Having postulated all that, I am fully aware that there are many people who are hesitant to agree upon the role of dietary amino acids as it relates to certain ailments. But, if you remember the postulates at the beginning of the chapter, i.e. formal 'mitosis' is a function of volume as it relates to greasy bubbles, i.e. cells, then illness based upon modified cellular cohesion seems much more realistic. Or in simpler terms, old theories of mitosis dictate that it is 100% controlled by DNA, which is an over-simplification. Continually growing cells require the EXPANSION of the cell to incorporate duplicates of all the components. Therefore, the rate of EXPANSION as it relates to the osmotic pressure variance, which brings in nutrients required for growth, is DEPENDENT on the surrounding environment. Or in scientific terms, a cell with a STATIC volume will undergo mitosis at a slower rate because: Doubling cellular components WITHOUT doubling the amount of water will force water out of the cell, i.e. a dehydration state, which scientists have already determined to be a factor that **inhibits** mitosis. (Also, it will take more time to bring nutrients into a cell when the volume is static because equilibrium of osmotic pressure slows things down dramatically.) All of which means the COHESION of cells is a factor in cellular mitosis. And since protein expression determines cellular cohesion, variable protein expression, i.e. cellular cohesion, is a factor to mitosis.

For example, let's hypothetically imagine that a cell deep within your liver doesn't get the lysine it requires to have normal cohesion with its neighbors. WITHOUT the volume regulatory imparted by the cohesion to the other cells, the cell is **able** to undergo mitosis to form two smaller cells. And as a result of two smaller cells, there is a variable amount of hydration about the proteins, which will modulate gene expression and possibly the proliferation rate. With a modified proliferation rate to produce smaller cells, that have their own specific cohesion amongst themselves, this new 'colony' is apt to form a tumor in your liver. In any event, cellular cohesion is amazingly important and varies based upon tissue, which has different amounts of vascularization and need for nutrients. All of which, is based upon one's diet and genetics.

Now I realize that I don't have the best diet, but I still want to help. In particular, I want people to know that their digestive tracks will NOT always work wonderfully. Butt, if you want to improve your chances of living for whatever makes you happy, then you might want to consider supplementing your diet with free amino acids...Especially the essential amino acids because they are VERY important in connective tissue, which is usually surrounded by fat and appalled by some diets. As for the diets with lots of fat, cholesterol is usually associated with the fat, i.e. it enhances cardiovascular problems in old people. In any event, free amino acids are cholesterol free, sugar free, fat free, and usually gluten free. The worst that could happen, if you drink enough water, is that you might build a little extra muscle such that you can shout at me for supporting God and evolution.

In conclusion, the decay of heavier elements on Earth has created a bimodal magnetosphere. And as a result of the magnetosphere and the continual vibration lighter elements via chemical bonds, Earth has evolved into an environment of lighter elements with extraordinary isotopic purities, which is conducive to life via NOT spewing out random neutrons/protons. Next, if Earth evolved via a Venus-esk atmosphere, then the decay of carbon dioxide into isocyanate is a viable method to the formation of basic heterocycles. (BTW, the subsequent decay of the nitrogen into carbon, within isocyanate, results in the formation of ketene, which crazy weird reactive.) Then, different transition metal nano-structures catalyzed the formation of single stranded monomers as well as served as a template to single stranded polymers. Then one day, a transition metal chelated by a single stranded polymer was surrounded by a grease bubble to create a system with a UNIQUE chemical environment that evolved based upon the trapping of substraights that diffused into the grease bubble. And finally, after millions of years, some beautifully complex shit happened and we appeared. Unfortunately, not enough of "us" have learned that supplementation with free amino acids might improve our lives.

Chapter 10: Green Tea

My Master's thesis sucked: I didn't publish a paper, I didn't find a new synthesis, and I definitely didn't make my advisor proud. (Sorry Dr. Kinstle.) In lieu of all those things, since my days at Bowling Green State University, I have learned a few things about LIFE, i.e. Multicellular organisms rely on communication across lipid bilayers, which is ALL based upon protein structure and the environment. Unfortunately, toxins can modulate protein structure, which is the reason why antioxidants are helpful to the body, i.e. they remove toxins. As for the method by which antioxidants remove toxins, it varies. Traditionally, antioxidants are defined as molecules that safely dispose of free radicals. All of which brings me back to my work at Bowling Green State University.

As a graduate student, I TRIED to synthesize an EGCG analog with enhanced solubility, i.e. positive and negative charges. Unfortunately, no matter how much I tried, I couldn't get the synthesis to work. But, what I did learn from these failures was that the carbons of trihydroxy-benzene were more nucleophilic than the oxygens...in some cases. In any event, to make a long story short, I eventually learned that this enhanced carbon nucleophilicity was the method by which antioxidant tannins were formed. All of which got me to thinking: Was it possible that neutral EGCG was diffusing into membranes and reacting with toxins via carbon nucleophilic attack to create excretable molecules?

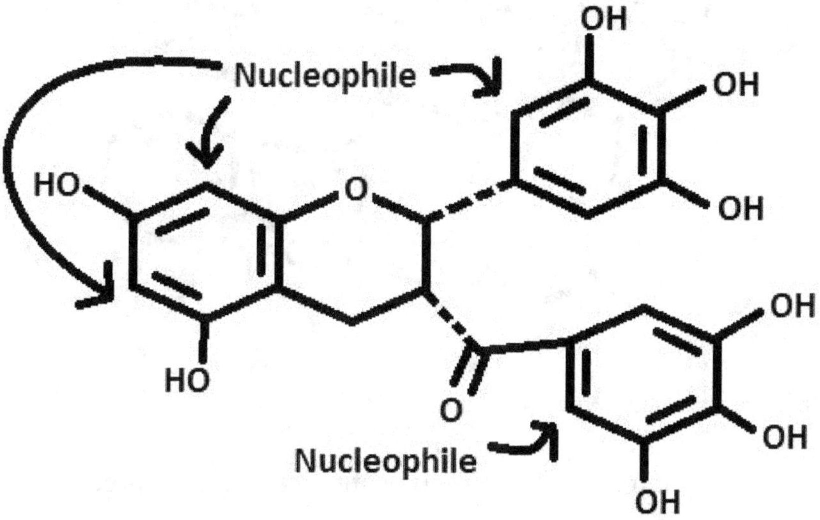

Figure 35: EGCG

As you can see in this figure, EGCG has six nucleophilic carbons that could react with toxins within the body. And just in case you were wondering, which I'm sure you were not, the hydrophobic environment within membranes will increase the nucleophilicity of the carbon atoms by decreasing the acidity of oxygen's hydrogens. (That was for all the nerds out there!) In any event, if my postulate about EGCG's activity within cellular membranes is correct, then the following EGCG analog might enhance EGCG's anti-toxin activity within cellular membranes.

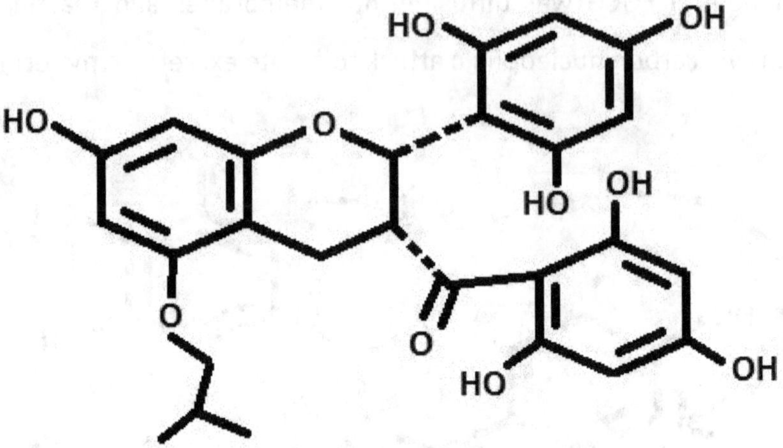

Figure 36: IDÉE FIXE POUR AIDER

In conclusion, intercellular communication can be disrupted by toxins within cellular membranes. Therefore, a slightly lipophilic EGCG analog might be better suited to remove toxins from cellular membranes such that the body my excrete them.

Chapter 11: How to survive torture

Hypothetically, let's say you want to torture someone because they have caused you emotional anguish. Unfortunately, you can go to jail for torturing someone...unless you're the US government. Therefore, you'll have to vent your anger via psychological torture. Granted, The Oxford Dictionary defines torture as "severe physical or mental suffering", but psychological torture has a much better ring to it. Now, is it torture when an acquaintance demeans you? Not really. Most people have low self-esteem, which requires them to perpetually demean others. But, when a group of people work together to demean you, this could be classified as torture. As to whether or not it is a hate crime, that's up to you to define...screw the Supreme Court and Dictionaries!

If I was going to psychologically torture someone, I would need unfettered access to information. Unfortunately, nobody wants to be Nixon. Therefore, the best route to gain unfettered access to an individual's life is by CLAIMING to be their protector. And if that doesn't work, then good old fashion blackmail and/or death threats will probably work. (BTW, it doesn't take much effort to get bad people to be worse people.) In any event, after gaining access, you need build a profile on the person you want to torture. Unfortunately, most people don't do a lot of journaling. That is why you need to get your hands on pictures. (This is why cops are always trying to spark suspect's memories by sliding pictures in front of them…at least on TV.) Or in simpler terms, someone has to TRAVEL around, interview acquaintances with the aid of pictures, and record all this information. But, random stories don't amount to psychological torture. For this, you'll need psychologist.

For the most part, people are predictable. From Presidents to gardeners, people want to be respected, not laughed at (unless they're a comedian), and/or be besmirched for their intellect or physical attributes. And on top of all that predictability, most people have trust issues, regrets, and/or questions that make their mind go ca-chunk. Therefore, the trick to psychologically torturing someone is to get their mind to whirl uncontrollably. And the best way to start this is: Isolation.

Once you have isolated someone, you can begin to whittle away their emotional defenses. And the best way to put a person on edge is to threaten them. From there, the evening news will probably give them a heart attack. But, if you have unlimited means and access, then you can craft commercials to exacerbate a person's particular fears. And if you're crazy enough, you can buy a telecommunications company and flash suggestive ideas in between commercials. (To the average person, it probably just looks like somebody in the control booth fucked-up and tried to slip in one more commercial.) Or, you could just create movies and/or TV shows that bring about extremely **relatable** stressful emotions. All of which, brings about one great truth: Stress is cumulative because most people can't compartmentalize. Therefore, continual psychological stress will eventually result in an emotional breakdown. Here are some common psychological STRESS venues:

1. Lost, unfaithful, unresponsive, etcetera, etcetera LOVE. (This usually only works for people of reproductive age.)
2. Self-worth, self-accomplishment, etcetera, etcetera. (This usually only works for NON-apathetic people.)
3. Illness, death, money. (This usually only works for people who have something worth living for.)

For example, you can torture someone sexually by saying they are inadequate, worthless, and/or undesirability, which is the reason why breakups are so shitty. Then, add in an extra-large scoop of complicated-LOVE and sprinkle in a threat of genital mutilation...and whala, you've got some pretty good psychological torture. Trust me, anyone who has ever been cheated on, dumped by a husband or wife, and/or contracted a STD from a monogamous 'love' interest, has experienced **extreme** psychological stress. Therefore, here are some typical thoughts you might want to put off pondering in the event you've identified someone is trying to psychologically torture you...outside a committed relationship.

1. Was my partner a pathological liar?

2. Did my partner love me?

3. Was my partner faithful?

4. Am I an adequate lover?

5. Has someone else been more sexually pleasing to my significant other?

Now for the most part, people avoid confrontation. (This might not be true, but let's just assume it for the moment.) Or in psychological terms, most people are willing to forgo a little psychological stress because prolonged confrontation will result in even greater amounts of psychological stress, i.e. people chose their battles. But, for whatever reason, some people have topics they're REALLY passionate about. And in the event you accidently push one or MORE of those hot button topics, you're bound to get some blowback. Fortunately for most Americans, this will only result in a terse exchange of words, i.e. freedom of speech. But, if this exchange involves a person/corporation with unlimited money and an average citizen, then prolonged psychological torture is inevitable. Therefore, think long and hard before you become a public figure with opinions that differ from people/corporations with unlimited money.

In the event, if you are the subject of psychological torture, i.e. you're a poor schmuck, you still have some options...although they're not great. And even though learning about the process by which someone will psychologically torture you is NOT fun, at least you'll be able to vaguely identify why you are stressed. All of which, might enable you to endure the psychological torture. In any event, here are some tips on how to avoid psychological torture.

1. Be apathetic.
2. Don't have an online profile devolving memories or pictures.
3. Don't date unless you plan to get married.
4. Save your money in a mattress.
5. Do NOT stand up for ANYTHING, i.e. NO whistleblowing. For example, just 'hmm' for help if you're getting raped.
6. Do NOT be political.
7. Do NOT support women's constitutional right to choice, vote, or get equal pay. (Wait, women still don't get equal pay?)
8. And above all, do NOT assume people are good.

Also, be very wary of hope, i.e. building a happy place. HOPE is a method to facilitate your continual observation/torture and it ADDS to your psychological stress when ideas are presented that are contrary to your HOPE. Why do you think there is a saying about 'dashing someone's hope on the rocks'? Quite simply, no matter how REMARKABLY similar names, events, and/or concepts might be to your past, remember they are simply trying to cause you psychological anguish. Accept the past is past and avoid wondering if your convictions were ill-conceived. Nothing is more torturous than doubt and what-if questions...just ask a Catholic.

In conclusion, it is possible you're crazy. Or, it is possible that someone is trying to invoke emotional stress in you by REPEATEDLY implying that you're useless, unloved, disposable, stupid, sexually incompetent, ugly, the product of some miracle drug, murder, rapist, pedophilia, and/or a gun smuggling military contractor that allowed DOR's Glock or MJB's AK47 to start the second war in Iraq. (The possibilities are literally endless.) By the way, a great way to continually invoke emotional stress is by associating things via implications. For example, continually associate a butt-pack with a Glock and pedophilia such that every time a person sees a butt-pack, they will get a weird feeling of psychological distress. In any event, to avoid psychological torture, remember to be hopeless, apathetic, and voiceless. Or, you could just use reverse psychology. And finally, save your money in a mattress because most politician/bankers will gamble your life savings away in the blink of an eye.

Chapter 12: Ethereal

Finding meaning in anything is a chore, especially when it comes to the minutia of science. God knows we've been fighting progress at every corner as it relates to the interpretation of words and sentences. In any event, I can only imagine how <u>difficult</u> it is for someone to visualize a **better** planet with LESS science. All of which, is the reason why I keep thinking and postulating...it's like an obsession or something.

In the event you are busy having a grand old life filled with family, holidays, work, and vacations, here is the gist of this book. (FYI, numbers don't correlate to chapters.)

1. Strong nuclear force = Neuproz groups that are exchanging negative Neuprotrons. (Strong nuclear force spectrum = [13N]>[11N]...[3N]>[N])
2. Weak nuclear force = Columbic integration between atomic particles based upon external electronics. (Weak nuclear force spectrum = {[13N]&[11N] > [11N]&[9N]...[3N]&[N] > [N]&(N)}.)
3. Each row of the periodic table contains a larger Neuproz grouping. The only caveat is that Noble gasses contain the Neuproz grouping from the previous row.

4. Helium decay from heavier elements is by the fission of four inclusionary neutron about the axis via the release two negative Neuprotrons. (Generally an axial decay, but circumference decay is possible with the appropriate alignment of extra neutrons.)

5. EVEN elements have more isotopes because the decay of larger Neuproz groupings into smaller Neuproz groupings is ALWAYS via the loss of 2 neutrons and 2 protons, which decay into neutrons via the release of 2 negative Neuprotrons. Or in simpler terms, the LOSS of 2 protons results in every other EVEN isotope. Also, symmetric isotopes are generally more abundant than NON-symmetric isotopes because of Earth's bimodal magnetosphere.

6. Atomic particles have a CORE of supercoiled interlocked electronic entities that have 7 layers/grooves, which correspond to the row of the periodic table and the Neuproz groupings. As a result of Layer 7 being closest to the CORE, i.e. [13N], the negative Neuprotron that grinds about this groove moves faster, exchanges faster between Neuproz particles, creates a stronger bond between the Neuproz particles, and creates a stronger columbic attraction between the [13N] particles. All of which, is the reason why smaller Neuproz groupings decay axially.

7. Even though the placement of extra neutrons about the Neuproz groupings creates multiple Stereo Isotopes within heavier elements and result in unique decay pathways, a general Isotopic Degradation

Pathway can be understood as a result of odd elements and isotopic abundances.

8. The continual degradation of the energy in matter results in atomic particles that are unable to forge larger Neuproz groups. Also, the decay of lighter elements in deep-dark-COLD positive space spews atomic particles randomly. As a result of this, pseudo-isotopes with inclusionary atomic particles are forged in space, which are more sensitive to an increase in negative energy. When pseudo-isotopes decay, they create semi-inclusionary magnetically active elements that are more apt to isotopically relax, thus spewing out more protons and creating more semi-inclusionary magnetically active elements. Additional factors of isotopically relaxing are: vibration, odd atomic orbitals, and the presence of bimodal magnetospheres.

9. The formation of bimodal magnetospheres is a function of non-magnetic entities having oval trajectories about stars, which results in repetitive heating and cooling. All of which, gives a method to the energetic alignment of dense COLD planetary cores, which create a bimodal magnetosphere via initiation, amplification, and alignment. As a result of current data though, it seems as if older planets WITHOUT dense cores can have synergistically aligned warmer cores. Also, there is a good possibility that planets are just the broken hearts of dead stars that have been swept up by Milky Way's swirl-ish nature.

10. As planets develop stronger magnetospheres, more elements isotopically relax, which results in atmospheres with greater amount of hydrogen gas. After cold dense planetary cores evolve into warm synergistically aligned planetary cores, the eventual cooling of the planet will result in no magnetosphere, i.e. Pluto.

11. People are uptight about evolution, but all the data and postulates points towards it...Sorry for the emotional anguish.

12. More people should go into Organic Chemistry.

13. Be a good slave and keep your mouth shut or you'll get a whipping.

References

1. *CRC Handbook of Chemistry and Physics*. Florida: CRC Press, 1994, 74th edition.

2. *The Cambridge planetary handbook*. Cambridge: Cambridge University Press, 2000.

3. *Pocket Tables: An Everyday Reference*. New York: Barnes & Noble, 2003.

4. *The Oxford: American Desk Dictionary and Thesaurus*. New York: Berkley Books, 2001, 2nd edition.